Wesley Earl Dunkle

Additional Titles

Published by the University of Alaska Press

Geology of Southeast Alaska: Rock and Ice in Motion
Harold H. Stowell

Rock Poker to Pay Dirt:
The History of Alaska's School of Mines and Its Successors

Leslie M. Noyes in collaboration with
Earl H. Beistline and Ernest N. Wolff

Innocents in the Arctic: The 1951 Spitsbergen Expedition
Colin Bull

With a Camera in My Hands: William O. Field, Pioneer Glaciologist
As told to C Suzanne Brown

Permafrost: A Guide to Frozen Ground in Transition
Neil Davis

Grewingk's Geology of Alaska and the Northwest Coast of America
Constantine Grewingk

Wesley Earl Dunkle

ALASKA'S FLYING MINER

CHARLES CALDWELL HAWLEY

UNIVERSITY OF ALASKA PRESS
FAIRBANKS

2006 paperback edition published by
University of Alaska Press
PO Box 756240
Fairbanks, Alaska 99775-6240

Printed in the United States of America

The paper used in this publication meets the minimum requirements of the American National Standard for Information Sciences—Permanence of Paper for Printed Library Materials. ANSI Z39.48–1992

Library of Congress Cataloging-in-Publication Data

Hawley, Charles Caldwell, 1950–
Wesley Earl Dunkle : Alaska's flying miner / by Charles Caldwell Hawley.
p. cm.
Reprint. originally published: Boulder, CO : University Press of Colorado, c2003.
Includes bibliographical references and index.
ISBN-13: 978-1-889963-93-8 (pbk. : alk. paper)
ISBN-10: 1-889963-93-3 (pbk. : alk. paper)

1. Dunkle, Wesley Earl. 2. Copper miners—Alaska—Biography. I. Title.
TN140.D82H38 2006
622'.343092–dc22
[B]

2006020332

Cover design by Dixon Jones, Rasmuson Library Graphics
Cover image of Curtiss-Wright airplane courtesy of Bob Repish, www.icarusbooks.com
Interior design by Daniel Pratt

In memory of
Bruce, Jeff, and Pete

All Dunkle Stout

Contents

Illustrations

Preface

My interest in Wesley Earl Dunkle began with the 1967 remark of a fellow U.S. Geological Survey (USGS) geologist, the late C. L. ("Pete") Sainsbury. Pete remembered a prospect in the Alaska Range called Golden Zone that he thought was worthy of evaluation for a new program on heavy metals, especially gold. Sainsbury—personable, tough, and frequently controversial—thought that the promoter of Golden Zone was a man named Earl Dunkle. The prospect proved to be of great interest and so did Dunkle. Over the next three decades, I found Dunkle's footprints throughout Alaska: on Chichagof Island, in the Brooks Range, at the Big Hurrah mine near Nome, and at the Golden Horn mine at Flat. His presence as a mining engineer and geologist was especially strong in the Willow Creek district of the Talkeetna Mountains, around Prince William Sound, at Kennicott* in the high Wrangell Mountains, and at the Golden Zone prospect. Further search established a Dunkle family heritage in the Allegheny plateaus of northwestern Pennsylvania and a mining heritage in Nevada and Africa.

During a fifty-year professional career, Dunkle set a pace in geology and mining that was difficult to follow. The pace accelerated in 1932 after Dunkle earned his pilot's license, and he barely slowed down until his death in 1957. Fortunately Dunkle and some of his friends and associates left enough of a record to allow me to reconstruct his life within the context of the era in which he lived.

---

*Kennecott Mines Company, established in 1906, took its name from Kennicott Glacier, in turn named for pioneer naturalist Robert Kennicott. *Kennicott* was misspelled in the company name. The spelling *Kennicott* is used herein for the town and mill site next to the glacier. *Kennecott* is used for the company or its mines.

Dunkle himself contemplated writing an Alaskan history centered on mining. He assembled material and, through a contact made by Mt. McKinley explorer Bradford Washburn, corresponded with a willing publisher. A coal project intervened, which left no time to write. In many respects, I have tried to tell the story that Dunkle wished to tell: the story about his life as a geologist and engineer, about Alaska and other mineral-rich places, about fascinating people, and about an America in love with technology and, in that simpler time, unafraid of growth.

Acknowledgments

MAJOR SOURCES OF MATERIAL ON DUNKLE included contemporary newspapers; hundreds of letters and mining project files; technical reports, especially Dunkle's 1954 report on the Copper River district; and a series of tapes recorded in 1953. Diane Brenner, then archivist at the Anchorage Museum of History and Art, provided a cassette copy of the tapes and pertinently suggested perusal of the 1915 issues of the *Knik News* and *Cook Inlet Pioneer* for news about Dunkle. Alaska historian Elizabeth A. Tower was able to identify several of the historical figures and events covered in the Dunkle tapes because of her own research on Stephen Birch and George Hazelet. My friend and geological colleague Donna Willoya made verbatim transcriptions of several tapes and assisted with others. Peggy Shedwick McCook of Dallas, Texas, furnished photographs of early Kennicott and allowed the use of her father's journal. Her father, William John Shedwick Jr., was a contemporary of Dunkle and Henry Watkins at Kennicott. His journal casts additional light on the life of young mining professionals during the birth of the Kennecott mines in Alaska.

Reconstructing Dunkle's life would have been impossible without the cooperation of the Dunkle family. Bruce Borthwick Dunkle, Earl Dunkle's youngest son, and his wife, Peggy, are the conservators of family papers and photographs. Dunkle's middle son, William E. (Bill) Dunkle, former pilot and executive with United Airlines, perceived his mother's loneliness during the long absences of her far-traveled husband. Bill was especially knowledgeable about his father's long involvement with Alaskan aviation. The oldest son, the late John Hull ("Jack") Dunkle, was able to fill critical gaps in the story because of the years that he spent in Scotland and England just before World War II. Harry and Gloria Bowman, longtime aviators and friends of all the Dunkles, coordinated my long conversations about Earl with Bill, Bruce, Peggy, Jack, and Jack's wife, Dorothy, in southern California in early 1997.

Edith Serkownek, executive director of the Warren County Historical Society in Warren, Pennsylvania, furnished information on Dunkle's early education, on the legal career of Dunkle's father, and on the history of the Dunkle family. I owe an immense writer's debt to the 1977 genealogy of the Dunkle family by Clara Adella Dresskell Page. In addition to her own detailed work, Mrs. Page reprinted stories written by Earl's uncle, Peter Snyder Dunkle: accounts that vividly reconstruct Dunkle family life in rural Pennsylvania in the mid-nineteenth century.

John Whitehead of Athens, Georgia (formerly of the University of Alaska and Yale University), suggested the use of Yale class histories to follow the career of Dunkle and his classmates. Claudia Arabasz, former librarian at Kennecott Copper Corporation in Salt Lake City, searched company files for Dunkle and furnished letters between Dunkle and Stephen Birch concerning the Mother Lode property. Kennecott Exploration Company allowed me to search files in Spokane and Anchorage. Geophysicist James Fueg and his wife, Esther, now of Anchorage but formerly of Cape Town, South Africa, read and suggested corrections to the chapter about Dunkle's life in Africa in 1929. James Fueg's mother, Mrs. Phyllis Fueg of Cape Town, researched parliamentary correspondence and college records of Billie Borthwick Dunkle in Africa. In the spring of 1998, James B. Barnes catalogued all the Kennecott field records on Alaska in Anchorage. Barnes flagged Dunkle material and set aside key references.

Randy Moore, formerly of Cambior USA, searched the Anaconda files at the University of Wyoming for correspondence between Dunkle and Anaconda's Francis Cameron and Harry Townsend. Mining scholar Mark Steen of Colorado furnished information on Dunkle's associate John Gordon ("Jack") Baragwanath and on a relative of Gladys Borthwick Dunkle, John David Borthwick. Archive librarian Sasha Stanley of the Cox Newspapers, Inc., Dayton, Ohio, found photographs of Harold E. Talbott after most avenues of search had been exhausted. Ms. Stanley was also familiar with a biography on Talbott's mother, Katherine, a work that supplied important information about Harold Talbott's early career.

Alec Jones of Vancouver, British Columbia, corrected geologic assumptions in an early draft manuscript. Jones worked in the Wrangell Mountains in the 1960s, searching for another Bonanza mine. He also introduced me to Ronald N. Simpson of Copper Center, Alaska, and Warren D. McCullough of Helena, Montana, who were experts on early-day Kennicott. Simpson had already written about Dunkle's role in the discovery and acquisition of the Mother Lode deposit at Kennicott. He also published a fictional account of the Kennicott story from his unique perspective as both Kennecott Copper buff and Ahtna native.[1] Lewis Green, who was Jones's associate in the Wrangell Mountains, also read the early manuscript and suggested changes from his perspective as a geologist-historian.[2] Other knowledgeable sources on the Wrangell Mountains—Donald H. Richter and E. M. MacKevett Jr., emeritus USGS geologists—reviewed the draft manuscript and suggested revisions.

Several historians aided in the research on Florence Hull and Gladys ("Billie") Dunkle. Tammy Marten, archivist at Oberlin College, Oberlin, Ohio; Susan Schwerer, special librarian in Sandusky, Ohio; and Robin Willoya, an Oberlin student from Anchorage, collected material on Florence Hull and the Hull family, which has long been affiliated with Oberlin. Dr. Norman H. Reid of St. Andrews University, Scotland, and Kathleen Clark of the University of Cape Town, South Africa, furnished information on the education and family of Gladys Borthwick Dunkle. Julie Poole, research genealogist at St. Andrews, helped resolve to the extent possible the ancestry of Gladys Dunkle. Poole located Gladys Dunkle's daughter, Diana Rimer Edwardes, MBE, in Kent, England. Mrs. Edwardes described the sometimes enigmatic life and career of her mother through correspondence and telephone conversation.

Former archivists at the University of Alaska–Fairbanks—Gretchen Lake and Sylvie Savage—helped in locating Kennecott-era files and photographs. Renee Blahuta, Fairbanks, and Bruce Merrill, archivist at the Anchorage municipal library, found material on a longtime Dunkle friend, Helen Van Campen. Historian C. Michael Brown of the U.S. Bureau of Land Management searched and annotated the *Fairbanks Daily News-Miner* for pertinent material on Dunkle. Through his colleague James Ducker, Brown furnished indices on all Dunkle items in the Fairbanks paper from 1920 through the 1940s. The articles allowed construction of parts of Dunkle's mining and aviation history that before were largely conjectural. Southeast Alaska historian Patricia Roppel flagged one memo in the Hirst-Chichagof file at the University of Alaska–Fairbanks that gave Dunkle's early 1920s salary and placed him at Contact, Nevada, in 1925.

Several Alaskans knew Dunkle and were willing to share their knowledge and insights and to suggest other leads to follow. This group includes the late Decema ("Dee") Kimball Andresen, the late Robert B. Atwood, the late Peter Bagoy Sr., Earl F. Beistline, Daniel Cuddy, Charles F. Herbert, the late Rodney L. Johnston, John Miscovich, John J. Mulligan, Jack Neubauer, the late Elmer E. Rasmuson, Mr. and Mrs. Frank Reed

Jr., and William A. Waugaman. The Anchorage Elks Club has maintained records on its members since its founding in 1915. Charles Johnson searched these records for information on Dunkle and his friends.

Luther ("Tex") Noey and Paul Ellis worked for Dunkle as young men at the Golden Zone mine. They furnished helpful information and some quite amusing stories about him. Anna Marie Till, widow of Dunkle's longtime mill operator, Leo J. Till, and her son, Vincent (Jerry) Herrold, shared reminiscences and photographs of life at the Golden Zone mine. David Kunzmann of New York sent photographs and accounts of his aunt, Charlotte Miller Stanford, widow of James Stanford. As a young engineer, James Stanford assisted Dunkle at the Lucky Shot and Golden Zone mines and helped to evaluate other projects. The Kirsch family—John; his mother, Rose; and his uncles, Andrew and David—spoke and wrote about Dunkle's last months in 1957 at Colorado Station on the Alaska Railroad.

Several people, especially Carol Young of Talkeetna and E. A. Holmberg of Colorado Station, shared their Dunkle letters and memorabilia. John Reeder, geologist and the geologic curator for the State of Alaska, furnished the Sydney Laurence painting, formerly owned by the Dunkles, which is used as the backdrop on the book cover. Dunkle had his greatest financial success at the Lucky Shot mine in Alaska's Willow Creek mining district. I owe Starkey A. Wilson and the late Tommy B. Medders Jr. for the opportunity to observe this mine. These men and Enserch Corporation entrusted me with supervising the Willow Creek project from 1979 to 1984. Any author on Alaska's Willow Creek district must acknowledge William M. Stoll's 1997 history of the district.[3] Stoll read the chapter on Lucky Shot, suggested corrections, and gave permission to quote from his book.

Several of Dunkle's projects were in or adjacent to national parks. A valuable source of Dunkle-related material is contained in the published reports by past and present National Park Service historians, including Geoffrey T. Bleakley, William E. Brown, Rolfe Buzzell, Logan Hovis, William R. Hunt, Frank Norris, R.L.S. Spude, and Melody Webb. Janet McCabe of the Anchorage office of the National Park Service encouraged me at an early stage and helped find materials on the closing days of the Kennecott mines in the Wrangell Mountains.

I am especially indebted to four people who agreed to act as continuing reviewers. They are Jo Antonson, state historian of Alaska; Lyle D. Perrigo, an engineering historian; Alden Todd, a widely published author; and the late William R. Wood, president emeritus of the University of Alaska. Freelance editors Lori Jo Oswald, Wasilla, Alaska; Sue

Mitchell, Fairbanks; and Alice Levine, Boulder, Colorado, helped to advance the manuscript at various stages. Acquisition editors at the University of Alaska Press (Pam Odom) and at the University Press of Colorado (Kerry Callahan) guided the general path of revision. Kerry, in particular, suggested numerous places where extraneous material could be cut and thus enabled the book to follow the life and mining career of Dunkle more directly. Evelyn M. VandenDolder is a wizard who is disguised as a copy editor.

Finally I wish to acknowledge help, encouragement, and support from my wife, Jenny Lind Hawley, and my sons, Andrew Bruce Hawley, David Lind Hawley, and William Theodore Hawley. Jenny is an excellent editor. More than anyone else, Jenny moved me as far as possible from my predisposition with geology, engineering, and technical writing to the more human side of this story. Andrew, a graphic artist, enhanced marginal illustrations and accurately reproduced good ones.

Introduction

WESLEY EARL DUNKLE TRAVELED THROUGHOUT ALASKA by foot, boat, and plane as a mining engineer and geologist from 1910 to 1957. During those years, he made many solo crossings of Alaska's frigid, silt-laden, glacial rivers: more than anyone should have attempted. He also took high risks in the air. An account of Dunkle's life, however, should not be simply one of danger and adventure. He chose the path of a professional miner at Yale's Sheffield Scientific School. Because mining and geology advanced more in his lifetime than during the preceding five millennia, the path was always upward.

Dunkle's life focused mainly on a complex experiment: the conversion of a natural object (a mineral deposit) into a fabricated one (a mine). A mineral deposit differs from common rock only in its degree of mineral concentration. If the mineral deposit is sufficiently rich, it is an ore deposit; if the ore deposit is developed, it is a mine. If Dunkle thought it necessary to found an airline or invent a process to aid the conversion, he did so.

Part of Dunkle's career was marked by romance and wealth, another part by loneliness and near poverty. Dunkle was not corrupted by wealth, however, and he smiled through hard times. A love of Alaska drove him forward. Dunkle never forgot the wild beauty he encountered when he entered Prince William Sound in 1910.

Abbreviations

AAS	Arthur A. Shonbeck
ADGGS	Alaska Division of Geological and Geophysical Surveys, Fairbanks
ADT	*Anchorage Daily Times*
AGDC	Anaconda Geological Document Collection, American Heritage Center, University of Wyoming, Laramie
AIME	American Institute of Mining Engineers (1871–1918) American Institute of Mining and Metallurgical Engineers (1919–1955) American Institute of Mining, Metallurgical, and Petroleum Engineers (1956–present)
APRD	Alaska and Polar Regions Department archives, Elmer E. Rasmuson Library, University of Alaska–Fairbanks (If photograph, citation also includes collection name and accession number.)
ASARCO	American Smelting and Refining Company
BBD	Bruce Borthwick Dunkle
BBN	Bert B. Nieding
CR & NW	Copper River and Northwestern Railroad
DF	Dunkle family, as memorabilia or photographs
DT: R5, t2, sA	Dunkle tapes; for example, reel 5, tape 2, side A. (There are twenty-seven cassettes that are equivalent to nine reel-to-reel tapes.)
ETS	E. T. Stannard
FDNM	*Fairbanks Daily News-Miner*
GBD	Gladys ("Billie") Rimer Dunkle née Borthwick
GC	Glenn Carrington
GPO	U.S. Government Printing Office, Washington, D.C.
GZM	Golden Zone Mine, Inc.
RFC	Reconstruction Finance Corporation
RRW	Ridgeway R. Wilson
USGS	U.S. Geological Survey
USSRM	U.S. Smelting, Refining, and Mining Exploration Company
WED	Wesley Earl Dunkle
WPA	Works Progress Administration

1

Heritage and Education

IN THE EARLY EIGHTEENTH CENTURY, the Dunkles and other German families left well-tended farms for the New World. A series of internecine wars gave them few options. If the men agreed to fight, death was a likely consequence. If they refused conscription, their farms were confiscated. Many of those farms had been passed from father to son for hundreds of years. Because of long-continued war, economic conditions throughout the region were dire, and most of the emigrants left for America with few possessions. Many had to come as indentured servants, who, by reason of their knowledge of farm and woodland, were welcomed as settlers as soon as they completed their bonded contracts.

Wesley Earl Dunkle, about whom this tale is written, descended from Dunkles who homesteaded in eastern Pennsylvania in the 1730s. They exemplified the knowledgeable, disciplined settlers sought by colonial visionary William Penn. Dunkle's father, named John Wesley by his staunch Methodist mother, left a near pioneer life to study law. John Wesley's son, called "Earl" by his family, was often sustained by the family's pioneer heritage during his long career as a geologist and mining engineer.[1]

The homestead where Earl's father, John Wesley Dunkle, grew up was chosen by Earl's great-grandfather, Michael, just before the War of 1812. Michael crossed the Allegheny Mountains in search of new land and found a pleasing tract bounded on three sides by a meander loop of the Clarion River (fig. 1). When John Wesley was a boy, his father, who was also named Michael, replaced a quickly constructed pioneer log dwelling with a roomy one-and-one-half-story home. Constructed from hand-hewn white oak, the Dunkles' new home was on an elevated site chosen for its panoramic view of the river. The construction of the new home, however, had been second in priority to that of a fifty- by ninety-foot barn with upper sills of pine that were ninety feet long without a splice. The sequence of building a barn and then a house was clearly established

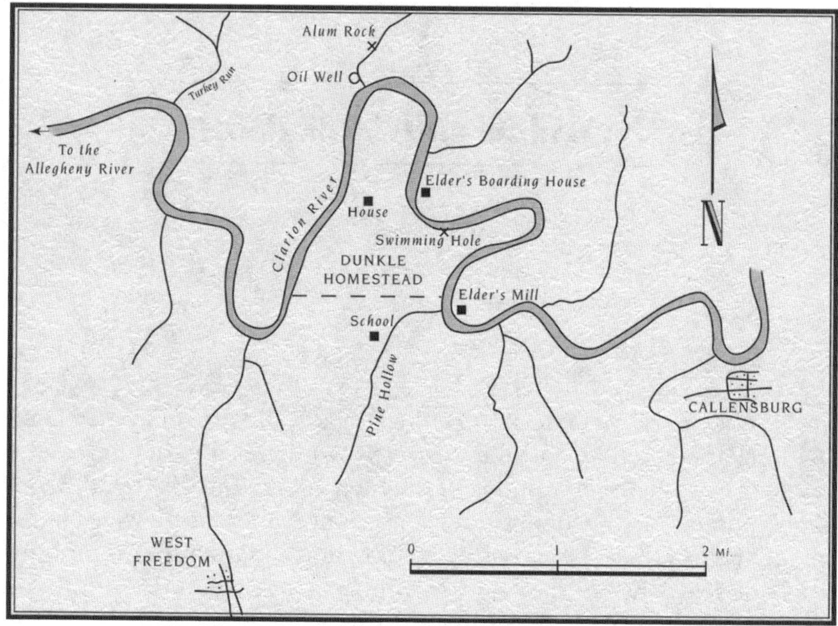

Figure 1. The Dunkle homestead on the Clarion River, northwestern Pennsylvania. The land was chosen by Earl Dunkle's great-grandfather, Michael Dunkle, at about the time of the War of 1812. Genealogy from Page, "'Dunkle'-Dusk-Twilight-Evening." Courtesy, Dunkle family.

in Pennsylvania Dutch country, where the saying was "[I]n five years a good barn will build a good house."

Michael Dunkle the younger married his neighbor, Caroline Boyer, an energetic lass nearly thirty years younger. An enlarged house was likely a necessity as seven young Dunkles entered the world and soon became the farm's main labor force. From their second birthday on, the Dunkle children were assigned tasks; the first was to supply the power for a six-foot-long cradle from an attached rocking chair. Later John Wesley's oldest brother, Peter, taught him how to use farm tools: "[H]oe, the axe, the scythe, and reins passed through my hands while I proudly gave instructions as to their use."

Caroline Dunkle brought religious conviction, youth, vigor, and a passionate belief in the value of education to her family. Her advocacy of education, for both sons and daughters, changed the lives of her children and their descendants. She was the first woman in her county and possibly in Pennsylvania to serve on a school board.[2] The preceding

Dunkles had been farmers or artisans, but of Caroline's six sons, only one—Henry—took up a trade. Three of Caroline's children, including John Wesley, graduated from a college or university.

Caroline probably also caused a change in Dunkle politics. Her husband was a Democrat until the Civil War, when he switched to Abraham Lincoln's Republican Party.[3] It was a choice of conviction, not necessity, because secessionists emerged among their neighbors, especially when war casualties began to mount. Caroline's oldest son, Peter, found Confederate sympathizers in the pulpit and at school. His grammar school teacher insisted on Southern loyalty, asking each day for a loud cry on which was the greater man: Abraham Lincoln or Jefferson Davis. The teacher's favorites answered loudly with the latter. Caroline visited the school to check on the teacher, and Peter escaped punishment when he explained, "Mother, I said Abraham Lincoln, so I did, but I was careful not to say it loud." All the Dunkle boys survived the Civil War, but many of their neighbors who served did not return. Hardest hit by the war were students from the nearby West Freedom Academy, who had joined en masse with their teacher to form the nucleus of the West Freedom Company.

The discovery of oil in northwestern Pennsylvania perhaps had a more long-lasting effect on the Dunkles than the Civil War. On 27 August 1859, four years after the birth of John Wesley Dunkle, Edwin L. Drake and his driller, Uncle Billy Smith, struck oil at Titusville, only a few tens of miles southwest of the Dunkle farm. Some of the Dunkles participated in the oil boom.[4] Men who would have stayed on the farm or developed an agrarian trade a decade earlier became entrepreneurs and began to travel. The Dunkles dispersed more in one generation than they had in the preceding century and a half.

The career of John Wesley Dunkle was influenced and later shaped by the oil industry. He worked his way through law school at the University of Michigan and began to practice law in newly prosperous Clarendon at the height of the Pennsylvania oil boom. (Oil had been struck at nearby Dutchman's Run in 1868.)[5] The assurance of a livelihood probably encouraged the courtship of John Wesley and Susan Amanda Dillon, the daughter of an Illinois minister. John and Susan were married on 19 September 1883 in Oil City, Pennsylvania. Their first son, Lester Dane Dunkle (called "Dane" by his family), arrived in 1885. Wesley Earl Dunkle was born on 4 March 1887 in Clarendon. The John Wesley Dunkle family was complete with the birth of a daughter, Murna Myrtle, about 1890.

John Wesley Dunkle won election as county attorney in 1887, and the family moved to the county seat at Warren, an oil-refining center on a

Figure 2. Downtown Warren, Pennsylvania, ca. 1900. Dunkle attended public schools in Warren, where his father served as county attorney. Courtesy, Warren County (Pennsylvania) Historical Society.

branch of the Pennsylvania Railroad (fig. 2). By the turn of the century, Warren was a busy and socially active city with clubs and lodges for almost everyone. It is still a small, prosperous city in a rural area in Allegheny National Forest.[6] At about age twelve, an alert and seemingly serious Earl Dunkle (fig. 3) must have known about several swimming holes, such as the one on his grandfather's farm. Many years later, when he constructed a freight scow on the tidal flats near Katalla, Alaska, Earl recalled that he had used the same built-up wood-canvas construction when he made canoes to explore the creeks of his boyhood surroundings. A near idyllic boyhood in Warren was assured by oil development that stabilized the economy of northwestern Pennsylvania, until then a backwater in a fast-growing state.

By the time Earl was in high school, he probably had developed a predisposition for an Alaskan career because of W. L. MacGowan, his high school principal and later the superintendent of schools in Warren. For many years, MacGowan guided vacation tours every June and July to "Yellowstone Park and the Klondike Gold Field of Central Alaska, Land of the Midnight Sun" and lectured widely about the north.[7] MacGowan and his faculty of ten strongly influenced Earl's public school education

Figure 3. Wesley Earl Dunkle ("Earl" to his family) at Warren, Pennsylvania, ca. 1900. Earl had two siblings: an older brother, L. D. (Dane), and a younger sister, Murna Myrtle. Courtesy, Dunkle family.

through the liberal arts and science classes they offered. Their college-bound students studied Latin, Greek, history, English, mathematics, and music. Texts for the science classes included Dana's *Geology*, Williams's *Chemistry*, and Avery's *Physics*. Earl probably took classes in plane and solid geometry.[8] Students who went beyond high school for further education were honored by being listed in "Our Pioneers." The "pioneers" named in 1906 included Earl Dunkle, his friend Will Dunham, and two Hue brothers at Yale University.[9]

Dunkle's future career demanded the physical strength that he inherited as a Pennsylvania Dunkle. Dunkle men typically were tall and powerful. The characteristics were so common that a Pennsylvania Dutch phrase, "Dunkle Stout-like," was applied generally, not just to the Dunkles, to describe extraordinary physical prowess. Earl Dunkle was no exception; he grew to be tall and lithe and was proud of his inherent strength and athleticism (fig. 4). His future, however, depended more on the knowledge that he would acquire in the rigorous academic program that he undertook at the Sheffield Scientific School of Yale University.

Dunkle is described as a mining engineer in most accounts, but some call him a geologist. He was both. During his first Alaskan job at the Beatson mine on Latouche Island, Dunkle was an assayer, surveyor, and all-around junior engineer. A few years later he was the mine superintendent. From 1912 to 1915 Dunkle had two jobs: mine geologist at Kennicott, Alaska, and field exploration engineer throughout Alaska and western Canada. As the field engineer, or scout, he examined mining

Figure 4. Wesley Earl Dunkle, ca. 1934. At age forty-seven, Dunkle carried 195 pounds on a slightly taller than six-foot frame. In Pennsylvania Dutch country, Dunkle men were noted for size and strength, so much so that big, strong men were called "Dunkle Stout-Like." Courtesy, Anchorage Museum of History and Art, Lu Liston collection, B89.16.

properties for Kennecott, occasionally finding one worthy of further exploration or development. He also followed the activities of competitive mining companies: Anaconda Company of Montana, the Alaska-

Gastineau and Bradley companies of Juneau, and Consolidated Mining Company of Canada.

During his highly productive years between 1912 and 1930, Dunkle is perhaps best described as an economic geologist, a term that he explained: "Economic geology means geology applied to mining. It is the study of the various geologic factors in and surrounding a mineral deposit so as to use these as a guide to the evaluation, development, and working of the mine."[10] The definition has breadth. The term *economic geologist* signifies one who has delved into the indirect factors that determine the success of a mine almost as much as the character and richness of the ore, factors that fall into the more limited province of the mine geologist. *Economic geologist* seems particularly appropriate for Yale man Dunkle. The journal *Economic Geology* and the professional Society of Economic Geologists were based at Yale for decades. The journal's first editor, John Duer Irving, began his career at Yale during Dunkle's junior year (1907), when he taught economic geology. Alan M. Bateman, who succeeded Irving as editor of the journal, worked with Dunkle in the Wrangell Mountains and along Prince William Sound as a consultant to Kennecott and taught economic geology at Yale after Irving's death.

Economic geologists must be generalists. Each mining prospect is analogous to a chain: the weakest link determines the strength of the whole. The first link is the amount and richness of the ore. If that link holds, one considers the next link: whether the material is amenable to concentration and recovery. Other factors must be answered from broader knowledge. Is the title to the property secure? Is the government that guaranteed title stable? Is process water available, both in quantity and by law? Are there indigenous people, and if so, are they friendly, healthy, and trainable or none of the above? Can the prospect be developed with existing transportation, or can a railroad be constructed, such as the one built in the Wrangell Mountains of Alaska through nearly two hundred miles of rugged terrane? A modern economic geologist would know something about reclamation practices and the protection of fisheries. Even in Dunkle's time, when there were few environmental regulations, certain factors had to be considered. By 1910 the time was long past when wastes from hydraulic mining could be discharged into the waters of California or other locales.

The importance of a broad educational background for those multiple tasks was endorsed by two of mining's most famous practitioners: John Hays Hammond and Ira B. Joralemon. Hammond's late-nineteenth-century course of study at Yale's Sheffield Scientific School included astronomy, botany, history, political economy, English, French, German, comparative philology, and the pertinent geological and engineering courses. Later Hammond even found his early Greek, which he had studied in preparation for Yale, valuable when he had to use the Cyrillic alphabet in Russia. Joralemon, who was a Harvard man, agreed: "In addition to the broader outlook . . . , work in many non-related fields helped me to analyze new problems with a more open mind . . . , to think instead of memorize."[11]

The education received by Dunkle and his contemporary Ivy League friends and associates during the first decade of the twentieth century was excellent. Classical preparation in at least Latin was still assumed, and the student had to master the essentials of economics, engineering, mathematics, chemistry, physics, and one or two modern languages.

Dunkle had the good fortune to enter college at this high point in geological education. Today's youth dream of becoming astronauts; in Dunkle's time, young men yearned to be like the engineers described in the novels of Richard Harding Davis (once the nation's leading journalist) and Mary Hallock Foote, who inspired Wallace Stegner's novel *Angle of Repose.*[12]

Earl Dunkle's choice of Sheffield Scientific School was also auspicious in its timing. For one or two decades, the Sheffield School was one of America's best schools of mining, largely because of the efforts of John Hays Hammond. Those efforts were built on a solid foundation that was laid in late 1847, when a scientific school opened as the School of Applied Chemistry in the new Department of Philosophy and Arts at Yale. Engineering and economics were early additions to the curriculum, but for years courses in mining were limited or dismissed as too lecture oriented and impractical. That image changed rapidly in 1903, when Hammond, the highest paid engineer in the world, began to fund the construction of buildings and laboratories, to redesign course offerings, and to import professors who had practical experience as well as theoretical knowledge in their fields.[13]

Because of Hammond's efforts and possibly his fame, mining enrollment at Yale rapidly increased. When Dunkle entered the Sheffield School in 1905, there were sixty-eight upperclassmen in mining and metallurgy. Fifty-one of Dunkle's freshman class of 380 men elected either mining or metallurgical options. The new program benefited from hiring men such as John Duer Irving, whose scientific collections, which he acquired through his work with the USGS, illustrated the rocks and ore deposits of the mining districts of Alaska, Colorado, Nevada, and Utah. Irving also collected publications from federal and state geological surveys, which then focused mainly on the western mining districts. Dunkle and his classmates thus had an excellent and up-to-date library.[14]

Dunkle's technical courses at Sheffield included mineralogy, physical geology, and analytical chemistry. He took William E. Ford's mineralogy class in the fall semester of 1907. Ford stressed the economics of valuable minerals but did not neglect classical crystallography. In his neat notes, Dunkle ordinarily followed his professor's lead but may have strayed regarding the rare mineral sperrylite, a platinum arsenide that had been found only at three localities. Dunkle thought that he could find another location and "get rich quick." Someone, perhaps Ford, penciled in a large blue question mark after Dunkle's remark. Ford also introduced his students to the statistics on mineral production. In 1907, Alaska, with its production from the newly discovered rich deposits of placer gold, was second only to Colorado in U.S. gold production. (Colorado had temporarily passed California and South Dakota because of the rich ores being mined at Cripple Creek.)[15]

Dunkle learned more "economic geology" in his physical geology class. The section on mineral veins in Le Conte's *Elements of Geology* was Dunkle's assignment for 26 November 1907. The text clearly distinguishes two types of veins: hypogene and supergene. The hypogene deposits were "primary" veins formed from metal-laden ascending hot water. The supergene deposits were "secondary," formed when cool oxygenated surface waters percolated downward and reacted with the hypogene deposits. Later at the Kennecott mines in Alaska, Dunkle used his knowledge of hypogene and supergene veins to resolve the origin of the rich copper mineral chalcocite.[16]

Dunkle's preparation in analytical chemistry evidently was also satisfactory. He worked briefly as an assayer at the Veteran mine in Ely, Nevada, and at the Beatson mine on Prince William Sound, Alaska. In remote Broad Pass, Alaska, Dunkle assayed his own samples because he was convinced that the assayers in the nearest commercial laboratories were salting their samples and reporting fraudulently high gold values.

Dunkle was a well-known member of his class. He served as vice president of his senior class, ranking along with scholar-athlete and president T. A. D. ("Tad") Jones (easily the most popular man at Sheffield) and class secretary Howard E. Church, who was elected as the man most likely to succeed. The miners, represented by Dunkle, seem generally to have done well. Out of approximately three hundred students in Dunkle's junior class (the second year in the three-year Sheffield program), thirty-one men, including Dunkle, earned general honors. Class editor Howard E. Church noted the large number of miners in the honors

Figure 5. Mining graduates, Sheffield Scientific School, Yale University, 1908. Dunkle is at the far left in the bottom row. His associate at the Kennecott mines, DeWitt Smith, is at the center of the third row. The man standing next to Dunkle might be David Irwin, another Kennecott associate. The three men met again in Africa in 1929. From Church, Class History. *Courtesy, Mary and Melvin Barry, private collection, Anchorage.*

group: Mining majors, including Dunkle, constituted slightly more than one-tenth of the junior class but took about one-third of the honors. In Dunkle's senior year, the miners did as well in the more prestigious elections to Sigma Xi, then the senior class equivalent to honors. Dunkle won election, as did future Kennecott colleagues DeWitt Smith and David D. Irwin. The Sheffield mining class of 1908 (fig. 5) set a record difficult to match.[17]

Dunkle also found time for social contacts as well as athletic training and contests. From 1906 to 1908, four men—Richard E. Needham, William E. Dunham, Howard Oliver, and Dunkle—organized an extracurricular decathlon during Easter vacations. In intercollegiate athletics, Dunkle's sport was rowing (crew). In his freshman year, Dunkle was one of four Sheffield men to row with the Yale University freshman class crew. He retained a dominant role in this strenuous sport during his last two years at Sheffield. The senior class yearbook had to be at the printers before the rowing season ended, but the editor confidently predicted that Dunkle, Laurence Ballard Robbins, and John Newton Peyton would again uphold the honor of Sheffield.[18]

Crew is undoubtedly one of the best sports for developing lung power and stamina. Because of the versatility of the decathlon events, the athletic discipline served Dunkle well during his later years in the wilds of Alaska. Some of Dunkle's Alaskan hikes of 1912 to 1924 are legendary. In 1914 Dunkle explored the newly found camp at Chisana (often pronounced "Shushana") and the surrounding country before he walked back to Kennicott. In total, he covered many hundreds of miles of extremely rugged Alaskan country cut by glaciers and broad silt-rich rivers. In 1924 Dunkle made a solo crossing of the Alaska Range through Anderson Pass and reached a destination on the railroad in less than twenty-four hours. Part of the same route, in reverse, had taken the pioneer Belmore Browne expedition several days to accomplish. In 1931 at age forty-four, Dunkle thought little of a twenty-five-mile snowshoe trip to the Lucky Shot mine, a trip so demanding that it almost finished New Yorker Jack Baragwanath. Perhaps Dunkle should not have expected too much from Baragwanath, who admitted that much of his training was with a bottle of Scotch or at parties with his socialite-artist wife. As late as 1940 Dunkle trotted easily across the wide, silty, and always unpredictable West Fork of the Chulitna River. In 1955, at age sixty-eight, he could still outwalk a strong, lean, young coal miner named William A. Waugaman. In the year 2002, octogenarian Waugaman could still outwalk many younger people.[19]

Dunkle made lifelong friends at Yale. Four men, the volunteer athletes of the decathlon—Earl, Will Dunham, Richard ("Peg") Needham (who were roommates and fraternity brothers), and Howard ("Parker") Oliver, probably Dunkle's best friend—remained close throughout their lives. Three other lifelong friends, all mining graduates of the class of 1908, were Henry Coffin Carlisle, Henry DeWitt (later just "DeWitt") Smith, and David Duryea Irwin. Dunkle worked with Smith and Irwin at the Kennecott mines in Alaska, visited their mining projects in Africa in 1929, and corresponded with them throughout his life. Henry Carlisle gave the necessary engineering approvals for Dunkle's Lucky Shot and Golden Zone projects in Alaska.[20]

Dunkle's college years closely coincided with a waning copper rush in Alaska. The rush began in about 1898 and accelerated with the discovery of the Bonanza lode in 1900. It lasted until 1907. Professor John Duer Irving had worked in the Wrangell Mountain copper field shortly after its discovery. His student, Earl Dunkle, wanted to learn more about Alaska and its bonanza copper deposits.

2

Alaska in the Good Years

ALASKA WAS OFTEN IN THE NEWS DURING DUNKLE'S FORMATIVE YEARS. In 1898 Lindblom, Lindeberg, and Brynteson, the "Lucky Swedes," discovered rich gold deposits at Nome, initiating the largest gold rush in Alaska's history. In 1902 an Italian immigrant, Felix Pedro, made another discovery in Alaska's interior near present-day Fairbanks. The rush that followed the discovery disappointed many prospectors who sought easier ground, but the Fairbanks district ultimately proved to be the richest gold area in Alaska.[1] Nearly synchronous discoveries of rich copper deposits initiated copper "rushes" and generated more publicity on Alaska. In the summer of 1900, two prospectors discovered the richest copper deposit ever found, the Bonanza in the high Wrangell Mountains. Stephen Birch, an engineer fifteen years older than Dunkle, bought the copper claims and secured the capital to develop the deposit.

Although the copper deposits of the Wrangell Mountains were phenomenally rich, they were worthless without steamships and almost two hundred miles of railroad. A young engineer, Wesley Earl Dunkle, who had been predisposed to Alaska since public school days in Warren, Pennsylvania, was as challenged by engineering and geological problems as he was by the search for gold or copper. When he arrived in Cordova, Alaska, on Prince William Sound in the summer of 1910, he found that construction crews had bridged the Copper River in a stretch between the Miles and Childs Glaciers, a feat that some men had deemed "impossible."[2] Within a year, the first load of copper ore was shipped from Kennicott.

An early Kennecott engineer, Lawrence A. Levensaler, later denied that Stephen Birch remarked on the occasion of the first shipment, "Well Louie, there goes the first load of ore from the Bonanza—where will we get another train load?"[3] It is true, however, that in April 1911, when Birch supposedly made the remark, the Bonanza had almost no ore developed for production, certainly not enough to justify the tremendous expenditures made on the mine. The ore in the Bonanza had to be developed by mine workings, and the ore in the Jumbo and Mother Lode, which sustained Kennecott's Alaska mines until 1938, had not yet been discovered. Earl Dunkle played a considerable role in those discoveries and in elucidating the geologic processes that led to the formation of the ore deposits.

As Dunkle came to realize, rich deposits of copper existed in the Wrangell Mountains because of events that began in the Triassic Period, about 200 million years earlier. Thousands of feet of basaltic lava that was somewhat enriched in copper spewed from cracks in the crust of the earth. As the volcanic activity waned, the land subsided below the sea, and as much as three thousand feet of limy mud covered the basalt flows. Heat, pressure, and concentrated remnant ocean brines gradually converted the mud to limestone and dolomite. The entire rock package, now called Wrangellia, was then carried north on an ancestral Pacific plate, perhaps for thousands of miles, and attached itself to the proto–North American continent. The exact migration history, which is still uncertain, would have been unknown to Dunkle. He did, however, hypothesize that 140 million years after the Triassic, during the Tertiary Period, copper was leached from the basaltic rock and was deposited along fissures in the overlying limy rock.[4] These fissure deposits became the great copper ore bodies of the Wrangell Mountains. The discovery of the first one, the Bonanza, in the summer of 1900 augmented copper discoveries made on Prince William Sound from 1897 to 1899. Collectively the discoveries fueled a copper rush that rivaled the gold stampedes to Nome and Fairbanks and, in the realm of national politics and the early American conservation movement, had consequences that were more lasting. The aftermath of the copper rush was still evident when Dunkle arrived in Alaska in 1910. He met and worked with many of the pioneers who started it.

Dunkle acquired further insight on the copper potential of the region from knowledge of the precontact copper mining done by the Ahtna and other Alaskan natives in the Wrangell Mountains and on Latouche Island—mining that, in fact, had initiated the white man's interest in the region. Three groups of natives had a history of copper mining: the Ahtna Indians of the Copper River region on the south flank of the Wrangell Mountains, the Tanana Indians on the north side, and the coastal Eskimos of nearby Prince William Sound. Forty years before Dunkle came to Alaska, explorer William H. Dall reported, "Native copper, occasionally associated with silver, has long been obtained from the natives of the

Átna or Copper River. . . . The original locality is unknown and carefully concealed by the natives, with whom it is an article of trade. The specimens have a worn appearance, as if from the bed of a stream. That this metal exists in large quantities in this vicinity, there can be no doubt." In 1885 Lt. Henry Allen's expedition confirmed the existence of copper deposits in a tributary to the Copper River in Ahtna territory. The tributary itself was a "copper river," the translation of the native name, which Dunkle pronounced as "Chi-tee'-nah." Allen called the Ahtna copper river the Chittyna; it is now spelled *Chitina.* The Ahtna's Chief Nikolai told Allen that their copper deposits were near the Chittystone (Chitistone), a tributary to the Chittyna. Natural copper polluted another tributary creek known as the Chittytoh (Chititu). Of the deeply yellow-colored Chittytoh, Nikolai told Allen, "copper gives [the stream] its peculiar color, and causes the water to be so distasteful to salmon that they never ascend the stream." In 1891 explorers Frederick Schwatka and Charles Willard Hayes located another copper deposit, which was long known to a Tanana Indian band, on the northeast flank of the Wrangell Mountains. Ancestral coastal Eskimos on Prince William Sound had known of many other deposits of native copper, but after the introduction of iron by seafaring traders and explorers, their needs for and knowledge of native copper had begun to fade. One of the nearly forgotten deposits was at the outcrop of the Beatson ore body on Latouche Island.[5]

The Alaskan copper and gold rushes were synchronous, but the former was more limited geographically. Except for activity on southern Prince of Wales Island in Alaska's Panhandle, the copper rush was confined to the coastal mountains that surround Prince William Sound and to the adjacent Wrangell Mountains. A few prospectors, such as John Bremner, who searched the Copper River area for copper and gold before 1885, anticipated the event. The copper rush, however, was most intense between 1898 and 1907, the same time that gold was discovered at Nome and Fairbanks. The same gold-rush-like hustle and bustle—of waiting, then stampeding—and the same gold-rush-like cast of characters—saints and scoundrels, honest laborers and dreamers—swamped Valdez and Cordova, the copper rush's entry towns on Prince William Sound.[6]

Fishermen had already discovered Prince William Sound. The ports were ice free and only a few days from Seattle by steam. Coastal estuaries teemed with salmon. Prospectors quickly found that the sound was also rich in minerals. In 1897 M. O. Gladhaugh and Chris Pederson staked the copper-rich Ellamar lode near the abandoned native village of Tatitlek on the coast. White prospectors living with the Chenega people on Latouche Island made other discoveries. One copper prospect, probably at Horseshoe Bay on the north side of the island, was discovered by a man digging for clams. The major copper deposit on Latouche, known as the Beatson or the Big Bonanza, was located on 7 July 1897 by A. K. Beatson, M. O. Gladhaugh, W. Ripstein, W. F. Heidern, and W. E. Hunt. A native led the prospectors to the massive copper-bearing cliff, which was about one-half mile from the shoreline. Beatson's interest in the location had been piqued by natives, who used the powdered copper ore (chalcopyrite) to produce the black stain on their bidarka paddles. Apparently unknown to the contemporary native people, the *gossan,* or oxidized cap (see glossary), that concealed the Beatson ore body also contained native copper. Earlier aboriginal people—those who lived in the area before iron was introduced—knew about the native copper and had mined it for many years. Evidence of their mining activities includes several wheelbarrow loads of stone hammers, many of which are broken, that were found buried in the soil below the copper-bearing cliff.[7]

With a few exceptions, most of the contemporary accounts of mineral discoveries in Alaska mention only the white prospector and give little credit to local natives. A notable exception is the account of mining engineer Francis Church Lincoln, who visited the Beatson mine on Latouche Island about ten years after its discovery. Lincoln credited the natives who led prospectors to both the Beatson and the Ellamar deposits. Furthermore, he seems to have had anthropological interest in the ancient mining of native copper at the Beatson site because he sketched one of the stone hammers found there. He also noted the ongoing mining by the natives, who used the rich ore as a pigment. Dunkle seems to have had similar interests. In his Copper River paper, he strives for the correct pronunciation of native words related to copper and is amused by the fact that white settlers had a Copper River with little copper, whereas the Ahtna had a real copper river, the Chitina. Although the copper deposits in the Wrangell Mountains that were hosted by basalt were of little commercial importance, Dunkle thought that Nikolai's vein, the first main location in the Wrangell Mountains, was the most important copper deposit hosted in the volcanic formation. As a wilderness traveler, Dunkle often acknowledged his debt to the Ahtna and other natives for their wilderness lore, such as how to establish comfortable camps, cross or navigate white-water streams, and prevent snow blindness. We know almost nothing about Dunkle's relationships with individual Copper River Indians, but he seems to have accepted them as worthy early prospectors and masters of their wilderness terrain.[8]

The Beatson mine, purchased by the Alaska Syndicate in about 1909, was Dunkle's first Alaskan project. Even before the Bonanza discovery high in the Wrangell Mountains, the Beatson was recognized as a major copper deposit. Geologist F. C. Schrader, who was attached to Capt. William R. Abercrombie's exploration expedition of 1898–1899, said the deposit was "phenomenally large." It was also rich. Samples from the Bonanza No. 98 and the Ripstein Ledge No. 140 claims assayed, respectively, 16.30 percent and 25.37 percent copper.[9]

Shortly after the 1897 discovery of the copper deposit on Latouche Island, prospectors flooded into Prince William Sound, but on an elusive search for gold. Encouraged by transport agents seeking northbound fares and drawn by the ill-founded dream of an easy route to interior gold fields in the Yukon region, thousands of men and women arrived. The earliest arrivals were there by February 1898. Of more than four thousand stampeders, the majority crossed Valdez Glacier into the Copper River drainage, but only with great difficulty. Gold, if it existed, did not pave the Copper River, as some of the stampeders had been assured. By May of the same year, a general exodus of gold seekers began. It was nearly complete by October. Some died on their epic hike. For those who remained, scurvy was the norm not the exception. About three hundred men wintered in the Copper River basin in 1898–1899. A few men, mostly Minnesotans associated with an experienced woodsman, R. F. McClellan, sought copper not gold. The group became Chittyna Exploration Company. In 1899 an Indian led Edward Gates of the Chittyna party to a vein of nearly solid bornite, the first rich find in the Wrangell Mountains. The prospectors named it the Nikolai in honor of the chief of the Ahtna Indians.[10]

The greatest discovery in the region was made on 22 July 1900 by Clarence Warner and "Tarantula Jack" Smith, prospectors who were loosely attached to McClellan's company. The men discovered the Bonanza, a solid mass of malachite-stained "glance," the common name for a steely-gray copper mineral. Later, perhaps on the same day, the rich outcrop was discovered independently by A. C. Spencer of the USGS.[11] Spencer called the miner's "glance" by its scientific name, *chalcocite.* The name was not important, but the fact that the chalcocite contained almost 80 percent copper was significant. A copper rush to the Wrangell Mountains was on.

By 1901 prospectors had found hundreds of copper deposits in the Wrangell Mountains in a northwesterly aligned belt more than fifty miles long. Most of the copper deposits were close to the contact between the ancient basalt and the overlying limestone, respectively the Nikolai

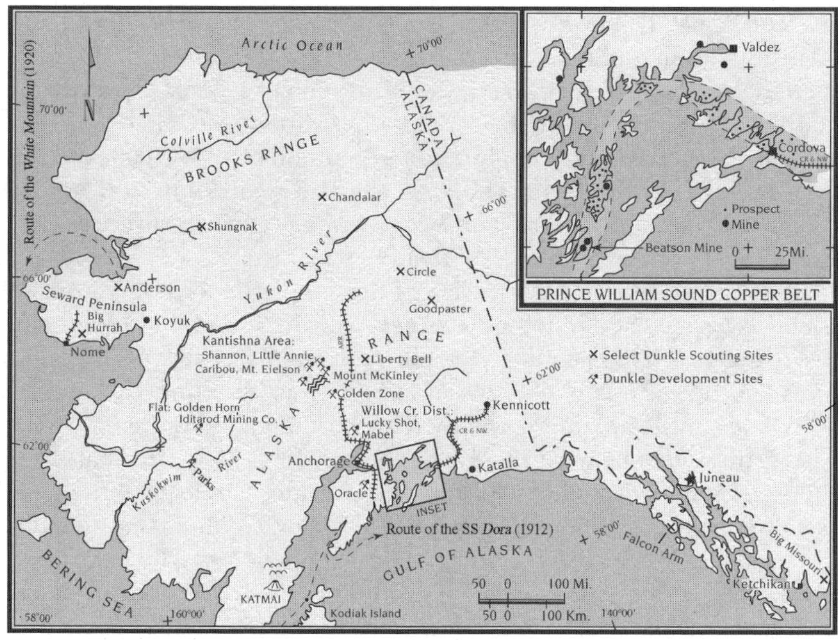

Figure 6. Dunkle country, Alaska. From 1912 through the 1930s, Dunkle explored Alaska from the southeastern panhandle north to the Brooks Range. His first Alaskan job (1910) was at the Beatson mine in Prince William Sound (inset). Author's compilation.

Greenstone and Chitistone Limestone. Copper prospectors also continued to comb the coastal lands and islands of Prince William Sound, especially during the period from 1903 to 1907. In 1908 USGS geologists U. S. Grant and D. F. Higgins Jr. identified 137 copper prospects and mines that formed a crude arc around the sound (see inset, fig. 6).[12] Of these, the Ellamar and the mines on Latouche Island were the most important.

After decreasing sharply in 1902, the price of copper rebounded from twelve to twenty cents per pound in 1907. In 1908, however, it slumped to thirteen cents. The copper rush, spurred by the discovery of the Bonanza, was over. Nevertheless, development continued at the Ellamar and Beatson mines and at a few other prospects; it accelerated at Cordova, the coastal access point to the Bonanza deposit in the Wrangell Mountains.

Several years had been consumed by claim acquisition and litigation over the ownership of the rich Bonanza discovery. The litigation was finally resolved in favor of Stephen Birch and his first backers, the sugar-rich Havemeyer family from New York. Birch and his associates had also obtained the backing of the house of J. P. Morgan, the premier transportation capitalists. The Morgan interests joined with the equally wealthy Guggenheims and the financiers Kuhn and Loeb to form the Alaska Syndicate, whose purpose was to gain steamship and railroad access to the remote deposits.[13] By 1908 intensive construction was underway on the Copper River and Northwestern (CR & NW) Railroad, which was to extend from Cordova up the Copper River to the Bonanza mine.

Concurrent discoveries of oil and coal increased development pressure on Prince William Sound and the coastal range. In 1901 a British consortium, the Alaska Steam Coal and Petroleum Syndicate, Ltd., developed a fifty-barrel-per-day oil well in Katalla Meadows. It was followed by several other oil wells. At times the coal of the Bering River field near Katalla appeared to be more important than either oil or copper. Railroad mania was in full swing. No fewer than four competing routes were proposed to connect coastal Alaska to the gold-rich Yukon basin.[14] Three of the rail routes started on the shore of Prince William Sound from the rival cities of Cordova, Katalla, and Valdez. Each route had been planned to develop the copper deposits on its way north.

Up until 1910 it seemed as if the Prince William Sound area would develop into a full-fledged industrial complex with mines, an oil field, and a smelter fueled with local coal. The oil field, however, failed to maintain its early promise, and the coal field was beset by legal problems that attained national prominence and forced adjustment of the earlier plans.

The seeds of the coal field's legal problems were sown in 1903, when Clarence Cunningham located and explored an extensive coal tract at Katalla. At that time there was no firm legal basis for locating coal lands in Alaska. After passage of a coal entry act for Alaska in 1904, Cunningham renewed his activity and applied for a patent on his claims. In doing so, Cunningham swore under oath that he and his associates would not aggregate more than 640 acres of claims and that the coal would be used by the claimants. In 1906 President Theodore Roosevelt withdrew all Alaska lands from coal development except those located since 1904, which protected the Cunningham claims. In 1907, however, the Cunningham group optioned more than five thousand acres of coal land to the Alaska Syndicate, thus violating legal restrictions on both acreage and coal usage.

In the initial years of the Cunningham case, Richard A. Ballinger was commissioner of the General Land Office, the forerunner to the

U.S. Bureau of Land Management. Subsequent protagonists in the Cunningham affair included a conservation-oriented investigator, Louis R. Glavis, and Gifford Pinchot, the chief forester of the United States. Glavis believed that there had been sufficient improprieties to cancel the Cunningham claims. In 1909 Ballinger was named Secretary of the Interior by President William Howard Taft. Recognizing a possible conflict from his earlier involvement in the Cunningham affair, Ballinger stepped aside and delegated adjudication of the case to Assistant Secretary of the Interior Pierce. Pinchot feared an Alaska Syndicate monopoly as well as the breakdown of the conservation policies that he had helped to establish during the previous Roosevelt administration. He urged Glavis to appeal directly to President Taft. Taft's investigation, however, exonerated Ballinger, and Taft urged that Glavis be dismissed; Ballinger complied.

Pinchot, bypassing the president, went to the media, which were then actively crusading against monopolies and trusts. A national magazine supported Pinchot to the extent that it alleged that an agreement existed between Taft and the Alaska Syndicate, an allegation that proved false. Taft had no choice but to fire Pinchot. Pinchot then went to past president Theodore Roosevelt and convinced him that Ballinger and Taft were out to scuttle Roosevelt's conservation policies. Roosevelt returned to the political fray when he organized the Bull Moose Party to oppose Taft's reelection. The votes that the Republican Party lost to Teddy Roosevelt's maverick party were sufficient to elect Democrat Woodrow Wilson president of the United States.

The controversy persisted for many years. In 1917 Daniel Guggenheim cornered Pinchot on his earlier charges of monopoly. Pinchot admitted that he could not justify his charges on the basis of fact, but he refused to apologize to Guggenheim.[15] The public, fed by the media onslaught, tended to perceive Ballinger and the Guggenheims as culpable. As noted by Alaskan historian Jeanette Nichols, "Public opinion was not greatly affected by the fact that the Syndicate's officials persistently denied such ownership and absolute proof of it was not adduced. For the practical purposes of politics, belief was as powerful as uncontroverted fact."[16] In 1940 an official investigation by Harold L. Ickes, Secretary of the Interior for Franklin Roosevelt, exonerated Ballinger, which formally closed the legal issues in the case but did not end the controversy.[17]

Exactly what young Yale graduate Earl Dunkle thought of the legal machinations of the Cunningham matter and the Alaska Syndicate is not a matter of record, but it can be reasonably inferred from his friendships and actions. Earl had met Stephen Birch a few days after his own arrival in Alaska in 1910. Birch impressed him then, and Dunkle's backing of Birch never wavered. Dunkle met Daniel Guggenheim in New York in 1915 and probably knew some of the Guggenheim family from corporate visits to New York that began in 1912. He regarded Birch, Murry Guggenheim, and J. P. Morgan Jr. as personal friends.[18] Almost certainly Dunkle believed in the rightness of the Alaska Syndicate in general and of Stephen Birch in particular.

In regard to the Katalla coal field, Dunkle probably shared the feelings of the citizens of Cordova, who essentially reenacted the Boston Tea Party when they dumped Canadian coal into the harbor because they could not mine or use their own. As a loyal company man guarding the syndicate's public image, Dunkle would not have gone as far as dumping coal or joining the citizens of Katalla, who hung Pinchot in effigy.[19] He was foremost an engineer and a builder. His main concerns would have been construction and production, and he would have disapproved of government actions that interfered with them.

As to mining on U.S. public lands, Dunkle and most miners of his time probably wondered why the government was entering a field that had been theirs for decades and why Alaska was being singled out. They had a point. The first national placer mining law was passed in 1870, more than twenty years after intensive placer mining had begun in California. Almost the same gap existed in Alaska, where the mining law for metallic minerals was applied in 1884, seventeen years after the purchase of the territory. As to coal law, Clarence Cunningham probably equated territorial Alaska with U.S. public lands that had been opened to coal entry since 1873. Basically Alaska passed a practical test: If it looked like vacant and unappropriated U.S. public land, then it probably was. The U.S. Congress, however, finally recognized the long-continued mining in trespass. If the trespassing miners were as large as the Alaska Syndicate, they were bound to be noticed. The syndicate also passed a practical test: It looked like a monopolistic trust.

The syndicate was on firmer ground with the copper claims. The mining law for metals was enforced in Alaska, and the key claims at Kennicott had been obtained by patent. The coal fiasco had two tangible results on the Alaskan economy. The CR & NW Railroad initially planned to fuel their locomotives with Katalla coal, and the syndicate had considered building a smelter—probably near Cordova—fueled by local coal. The legal issues raised at the Katalla field clouded the development of Alaskan coal for many years. The syndicate therefore decided to power its steam engines with fuel oil and to use the existing Guggenheim smelter at Tacoma, Washington, instead of building one on Prince William Sound. The impact on citizens of Alaska was more extreme. Thousands of people who had bet their lives and fortunes on Alaskan coal left the territory after 1910.[20]

Regardless of the fuel issue, the mines in the Wrangell Mountains could not be developed without a railroad. By the spring of 1908, Dunkle's last college semester and two years before he moved to Alaska, some of the railroad groups had lost heart. The Alaska Syndicate had won the battles of Keystone and Abercrombie Canyons against rival railroad constructors, but its actions had alienated many potential supporters. After a destructive storm wiped out the tracks and docks at Katalla—the coal and oil town and the proposed origin of one railway—the Cordova-based Copper River route triumphed over all competition. A few men still asserted, however, that the Copper River could not be conquered by rail. They also said that the Alaska Syndicate did not really want access to their rich Bonanza claims. One of those men was Herman Barring, who represented a rival railroad based in Valdez, Alaska. Barring told the chairman of the U.S. House Committee on Transportation, "Any competent railroad engineer in the country will stake his reputation on the impossibility to build a railroad up the Copper River, for the reason that there are glaciers there 2 and 3 miles wide on both banks of the river." Barring also told the committee that the reason the Alaska Syndicate wanted to move the rails to such a god-awful place was "to stop it [copper production from the Bonanza] so that it would not break the price. . . . These people who are opposing us are the king bees in copper. They do not want the price to go down." Mr. Barring was even more definite in his opinion of Stephen Birch, the engineer-founder of the Kennecott enterprise. "In my opinion the fact that this man is still alive is the best proof of the law-abiding character of the people of Alaska."[21] The natural obstacles were almost as formidable as Barring described. As he spoke, however, more than three thousand men were assembled in Cordova to show the skeptics that the syndicate was going to build a railroad up the Copper River and mine the Bonanza ore, almost without regard to the cost.

In August 1908, when young college graduate Earl Dunkle had just arrived at an iron mine in Coleraine, Minnesota, supplies were moving up the CR & NW Railroad across a temporary bridge at Mile 27 east of Cordova. The first load of supplies crossed the Copper River on the Fourth of July. By 5 October, when Earl was sampling iron ore at Coleraine, the end of the tracks was at Mile 49, the starting point of the Million Dollar Bridge. The bridge would cross the Copper River between Miles and Childs Glaciers, even if men had to whittle the advancing ice by hand.[22]

To Dunkle, descriptions of Alaska's construction projects and future mines were as exciting as fiction. Although he thought that his work at the iron mine was interesting, it was tame compared with the image of Alaska painted in newspapers and national periodicals. Dunkle vowed to make the Copper River country his destination. Other opportunities also existed in the north. Since 1898 a man could stake a homestead in Alaska. If Alaskan pioneers had their way, development initiated by fish, fur, and minerals would progress to farms, cities, and permanent civilization in the north. Some of the transportation barriers were already falling. Maj. Wilds Richardson's Alaska Road Commission, established in 1905, had begun to convert Capt. William R. Abercrombie's old telegraph trail into a road, all the way from Valdez to Fairbanks. Judge James Wickersham, in his first years as Alaska's nonvoting delegate to the U.S. Congress, would not give up until Alaska was a territory with a home rule legislature.[23] Even if the judge was prepared to fight Alaska's mining barons, who had spurned his advances toward representation, thus beginning a decades-long enmity, there should still be room in Alaska for an ambitious young engineer.

The first decade of the twentieth century has been called the "Cocksure Era."[24] The term is an exaggeration when applied to Dunkle and his classmates from Sheffield School. These men were certain of their course in life; they were confident but not overconfident.[25] They were prepared for an era when America and American technology began an explosive period of growth. Electrification of the nation demanded more copper than the older mines could produce and more gold to pay for capital improvements. American science, the basis of technology, set a daunting pace. As the first decade of the century closed, nothing as yet had signaled the end of this exciting period of invention and possibility.

Earl Dunkle was a quintessential product of this unique time. One American writer, Walter Lord, called the period (1900–1914) the "Good Years." In Europe it was the apogee of a decades-long interval between destructive wars, La Belle Époque. One self-styled "professional observer," a sometime mining engineer named Herbert Hoover, thought that the first decade of the new century was part of the "happiest period of all humanity in the Western World in ten centuries."[26]

In North America gold seemed to initiate the good years. The discovery of the bright yellow metal in the Klondike in 1896 and at Nome in 1898 helped to dispel the gloom and pessimism that attended the Panic of 1893, when silver prices plummeted. After that financial free fall, the nation, weakened by decades-long deflation, plunged into depression. Based on the gold discoveries in the north and even greater ones in

South Africa, it appeared that there was no shortage of gold in the world.[27] Gold was widely adopted as a monetary standard. In the United States, the archaic battle over bimetallism—gold versus silver—which for decades had divided rich from populist America, was effectively over by 1900.

More than monetary policy changed. Social muckrakers and business trust busters were in the news and sometimes in elective office. Curbs on business excess, rapidly changing technology, a new freedom in the arts and social fabric, and a rapidly developing conservation ethic characterized the period from 1900 to 1914, when Dunkle was at high school, college, and the beginning of his mining career. The new conservation ethic certainly played a part in the controversy surrounding the Katalla coal field in Alaska. Some of that conservational future was reflected in Dunkle's 1908 class at Sheffield School, where Dunkle and thirty-five other men graduated in the mining and metallurgical programs. Tension between the development and conservation viewpoints was evident in Yale's active forestry school, which graduated eight men in 1908. Graduates Edmund Prouty, Medary Stark, and perhaps Meyer Wolff anticipated a career in commercial forestry, but four others—Will Dunham, Henry Humphrey, Aldo Leopold, and Abbot Beecher Silva—looked toward conserving the resource with careers in the newly formed U.S. Forest Service. The four government-bound foresters were at that time clearly in the minority at Sheffield. The constructors (miners and engineers)—more than two-thirds of the 1908 Sheffield class—reflected the times.[28] Business was still America's ship of state, and the country was ready to grow with only minor trimming of its sails.

The dominant view of Dunkle's generation reflected a general optimism and the belief that technology could continue to transform and improve America. There were grounds to justify the faith in technology, especially if it was linked to a laissez faire economy. By 1900 the United States was the world's leading industrial power. There were eight thousand automobiles on the roads, some of which were paved. The northeastern United States and larger cities across the nation were lit by electricity.[29] The rate of technological advance seemed to increase exponentially. In 1901 Marconi sent the first transoceanic wireless message, the letter *S*, and radio was born. In 1903 the Wright brothers unlocked the secret of mechanically assisted flight, a secret that later intrigued Dunkle. From the 1890s to the 1920s, men such as Henry Ford, Elwood Haynes, R. E. Olds, and Frank Duryea converted the horseless carriage into an almost modern automobile. Thanks to Henry Ford's innovations, the automobile could be mass-produced, and Ford's workers were paid well enough to buy one, causing a social revolution. The first decade of the century saw the first practical uses of x-radiation, the isolation of radium, the first modern motion-picture process, the first modern motion-picture theater, and improvements in the phonograph that, in 1909, allowed engineers at remote Kennicott, Alaska, to listen to recordings of *The Merry Widow.*

Mine and mill technology also advanced explosively, both figuratively and literally. Compressed air or steam, instead of men with sledgehammers, drove drills into solid rock. The rocks were broken by high explosives instead of black powder or fire setting. The stamp mill, which had been the standard for crushing ore since Agricola's time in the early 1500s, yielded to jaw and gyratory crushers. In Alaska Frederick Worthen Bradley applied technology to all aspects of mining and milling. By 1900 Bradley's Treadwell mine near Juneau was the world's largest underground gold mine, producing more than $1 million per year from rock that assayed only about 0.1 ounce of gold per ton.[30] In Minnesota iron was mined in large pits by steam shovels that loaded strings of steam-drawn rail cars.

Many of mining's innovations were driven by an almost insatiable demand for copper, a demand spurred by increased electrical use in industry and at home. The rate of increase can be tracked by the nation's production of the metal: in 1850, only 700 tons; in 1870, 14,000 tons; in 1890, 130,000 tons; and in 1910, 540,000 tons, or more than 700 times as much copper as was produced in 1850.[31] One miner, D. C. Jackling, thought that he could solve the copper supply problem, and by 1905 he was ready to try at Bingham, Utah. Jackling proposed to mine low-grade copper ore with the same open-pit technology that was used in the iron ranges of Minnesota and Michigan. He soon demonstrated that it was possible to mine material containing only 2 percent copper, whereas underground miners were hard-pressed to mine ore that contained less than 10 percent copper.

The development of earth-moving technology and the low unit cost of operation are rightly credited for Jackling's success at Bingham.[32] There was, however, also a geologic reason. A very low-grade copper deposit that was formed tens of millions of years before had been upgraded by the process of *supergene enrichment* (see glossary) during relatively recent geologic time. The theory of supergene enrichment was new in the early 1900s; it was soon applied at Bingham, Utah; Butte, Montana; Ely, Nevada; and other locations.

Supergene enrichment is initiated by oxygen-bearing surface waters that react with pyrite and other, often more valuable, sulfide minerals contained in the primary ore. The main products of the reaction are sulfate acids that dissolve copper and other soluble metals. At Bingham acidic surface waters leached the copper and percolated downward to the water table, which was about one hundred feet below the surface. As oxygen was depleted and the acids were neutralized below the water table, the copper reprecipitated. Gradually nearly all of the copper above the water table was leached, leaving a multihued reddish mass of rock rich in iron and silica called the *cap* or *capping*. Below the capping a blanket of enriched copper ore about eight hundred feet thick graded downward into the primary ore, which was mostly granite that contained a few percent pyrite plus valuable copper and molybdenum minerals. This *primary ore* contained about 1 percent copper, a grade that was unmineable in Jackling's time.

The primary ore looked basically like a granite porphyry with sparkling metallic crystals; hence its name, *porphyry copper*. The supergene ore looked even less prepossessing. It was a clay-rich, almost dead-white rock with pepperlike flecks of the sulfide mineral chalcocite. Each fleck of chalcocite, however, contained almost 80 percent copper. A rock with only about 3 percent chalcocite thus contained slightly more than 2 percent copper. That was Jackling's ore: a supergene-enriched copper porphyry.

Most miners, at least the vocal ones, thought that Jackling was crazy. He proposed to mine at the rate of many thousands of tons per day instead of the fifty or few hundred that were produced at underground mines. The skeptics thought that Jackling's big mine would just lose money. According to Ira Joralemon, the prestigious *Engineering and Mining Journal* editorialized, "On the Company's own showing, the more ore it has of the kind it claims, the poorer it is." The skeptics had not reckoned in the lower per-unit cost of the entire operation, but Jackling had.[33]

Jackling's success at Bingham ushered in the extensive open-pit mining of low-grade ores. It supplied the world with copper and a problem: where to put the tailings. This dilemma still exists today.

Dunkle was an educated observer and at times a participant in mining's transitional years. He was introduced to the theory of supergene enrichment in 1907 in his physical geology class at Sheffield, where he studied how the supergene process had worked at Ducktown, Tennessee.[34] In 1909, as a smelter assistant and assayer, Dunkle observed the open-pit mining of Bingham-like supergene chalcocite ore at Ely, Nevada. In 1912 at Kennicott, Alaska, Dunkle worked in the richest chalcocite mine ever discovered. There he proposed that the chalcocite had formed from ascending *(hypogene)* solutions. Hypogene theory was even newer than supergene theory, but Dunkle had also been exposed to it at Sheffield.

The theory was strongly advocated by one of America's leading economic geologists, Waldemar Lindgren.[35] Lindgren was convinced that veins and most *primary deposits* had formed as hot solutions ascended from deep sources toward the surface of the earth.

Dunkle was also in a position to observe the later stages of the formation of a mining dynasty. The early part of the twentieth century was certainly the Golden Age of the Guggenheims, Dunkle's ultimate employer from 1910 to 1930. During that period, the Guggenheims' transition from lace merchants to mining kings was completed. Dunkle was also near the geographic limit of the concurrent cultural revolution, which in Europe was a significant part of La Belle Époque. The era had an effect even in remote Alaska, thanks to the stampedes for Alaskan metals. The gold and copper rushes agglomerated a variety of men and women.[36] In addition to the working prospectors and newly emigrant miners, there were bankers, lawyers, con men, surveyors, prostitutes, teachers, preachers, and a few artists, writers, and musicians. In this latter group, Alaska harbored no Picasso, Twain, or Stravinsky, but the harsh new territory attracted and held men and women with sufficient talent to record Alaska's natural beauty and a raw new civilization. A few of them had participated in the broader world of art, literature, and music before they came to Alaska.

Earl Dunkle knew several of these exceptional people. From his extensive remarks in the tapes that he recorded in 1953, it is apparent that he found them fascinating and worthy of historical appreciation. Two of them, writer Helen Van Campen and musician Willis Nowell, were among his closest friends. Dunkle also knew artists Sydney Laurence and Eustace Ziegler and novelist Rex Beach, probably very well. Artists Laurence and Ziegler have retained their reputations. Beach is interesting to Alaskan history buffs. Van Campen and Nowell are now virtually unknown, although they were important in their day. Critics ranging from *The Bookman* before 1910 to George Jean Nathan in the 1930s agreed that Helen Van Campen, once the wife of Dunkle's boss at the Beatson mine, could write very well. In particular, *The Bookman* wondered how a woman could know all the things that Helen knew about New York lowlifes.[37] Nowell—first a classically trained violinist, then a prospector and agent for Alaska Steamship Company—told Dunkle how he had successfully competed with famous American violinist Maude Powell in Paris and how he had once refused a command performance before actress Sarah Bernhardt, also in Paris. Dunkle believed that Nowell had studied with "J" (probably master violinist Joseph Joachim in Berlin) and that Nowell had owned two Stradivarius violins.[38]

Sydney Laurence painted *Hinchinbrook Island* (the background for this book cover) for Dunkle. Dunkle and artist Eustace Ziegler also must have crossed paths many times. Ziegler's territory as an Episcopalian missionary—Cordova to Kennicott—almost duplicated Dunkle's initial scouting territory for the Alaska Syndicate. Ziegler's artistic career was closely tied to Kennecott Copper Corporation through the timely purchases made by E. Tappen Stannard.[39] In *The Iron Trail,* romantic novelist Rex Beach used almost total reportorial accuracy to recount the construction methods and horrendous physical problems faced by the builders of the Million Dollar Bridge on the CR & NW Railroad east of Cordova.[40]

Thanks to Edison's recently improved phonograph, some of the culture of the era reached the engineers at Kennicott. In 1909 an editor for *Colliers* brought new recordings of *The Merry Widow* to Kennicott, where they were played on a Victrola given to Stephen Birch by his patron, Mrs. Theodore Havemeyer. The operetta "played" in Kennicott in the same year that it was first performed in Paris.[41] In 1953 Dunkle's friend Henry Watkins could still whistle tunes from violinist Maude Powell's early recordings, which he had heard in Kennicott before 1915.

The Kennecott mines, however, were more affected by technological changes and the emerging conservation movement than by the culture of La Belle Époque. Antitrust fervor painted the Alaska Syndicate of J. P. Morgan and the Guggenheims much grayer than it deserved. In so doing, it put a damper on the development of Alaska.[42] Coal fields that might have been developed and would have employed hundreds could not be developed later. The rich but small and discontinuous coal deposits were incompatible with large-scale development, which was the later alternative. Railroads that could have been built were not. The copper mines held on but ultimately could not compete with the huge, low-grade copper deposits found elsewhere, and they too were finally closed.

3

Apprentice, Scout, and Economic Geologist

Dunkle's sequence of academic education followed by practical work in the mines was a standard pathway to professional mining careers until nearly the present day. It was a practice carried forward from the Royal Mining School at Freiberg in northeastern Germany, where young engineers attended classes in the morning and worked in the mines and mills of the historic mining district in the afternoon. In the United States, Anaconda Company led the industry when, in 1900, Horace V. Winchell engaged the first professional staff of geologists.[1] For years Anaconda put newly hired geologists and engineers underground with an experienced miner for a several-month training course in practical mining to complete their education.

Dunkle's intensive postcollege apprenticeship in geology and mining lasted from August 1908 until late 1911. During this time, Dunkle held entry-level positions in mines, mills, and smelters in Minnesota and Nevada and on Latouche Island, Alaska. After completing this field apprenticeship, Dunkle pursued a mining career for almost fifty years. At times his career returned him to Nevada or brought him to British Columbia, the Pacific Northwest, and even Africa. His first reasoned choice, however, led him to Alaska, and that is where he pursued most of his career. Dunkle later reminisced that in the 1930s he probably knew physical Alaska about as well as anyone. In that period, it might have been called "Dunkle country" (fig. 6).

At the beginning of his apprenticeship in late June 1908, Dunkle may still have been loosely tied to Yale. He was at the crew training camp at Gales Ferry, Connecticut, where he received some tongue-in-cheek encouragement from his brother. Dane was in New York and sent Earl a postcard of the Nathan Hale statue in City Park: "Be good & you may have one of these some day," signed brother Dane.[2] By early August, Earl was at the Canisteo open-pit iron mine in Coleraine, Minnesota. At

Coleraine Earl began a practice of informing family members of his whereabouts and projects through letters and postcards:

> Dear Father,
>
> I received the books and your letter a few days ago. Let me answer a few of your questions. Coleraine has a population of about 600 or 700. . . . We are about 85 miles from Duluth, on a branch of the Duluth, Mesabi, and Northern. . . . The railroad tracks lead out to the slushing dump. . . . [A]ll this ore up here lies under a bed of glacial gravel which is from 30 to 80 feet thick and must be removed before the ore can be got at. . . . Last month over 400,000 cubic yards of sand and gravel was removed from these pits. This will some day be the largest open pit iron mine in the world. . . . I like my work real well. I get a change tomorrow and hereafter will take samples of the ore and various products of the mill so that they can be assayed. My love to all.[3]

It was at Coleraine that Earl Dunkle read of the Alaska Syndicate's efforts to reach the Bonanza copper deposit in the Wrangell Mountains. He decided to go there. He made an interim stop at Ely, Nevada, perhaps for convenience or to gain further mining experience. Dunkle's position at Coleraine had been obtained through J. F. Cole, ex-president of Oliver Iron Company.[4] Cole may have aided Dunkle's move to Ely since he also controlled Giroux Consolidated, the owner of the Kimberley copper deposit at Ely.

Dunkle's first job in Ely was at the McGill smelter. In contrast to a rather lengthy letter to his father (on the backs of six postcards) about the Minnesota iron mine, Earl used far fewer words to his mother in describing his new job location. Three postcards with one line each succinctly cover the trip from mine to smelter:

> This is where our ore comes from.
> Halfway to the smelter.
> And this is where it comes to.[5]

At Ely Dunkle observed the second open-pit copper mine in the United States (the first was at Bingham, Utah). Dunkle described the pit: "This is Copper Flat the biggest ore deposit in the district. It is about 2 miles east of us and 18 miles from the smelter. . . . The pure white stuff is ore and the darker stuff on top is stripped off and put on a dump. The ore will run about 2.4% Copper."[6] Dunkle also made a contact at Ely that was critical to his later career. He met W. H. Seagrave, the manager of the underground Veteran mine (fig. 7), and rustled the job as assayer.

Figure 7. Open pit (top) *and underground* (bottom) *copper mines, Ely, Nevada, 1909. On the top postcard, Dunkle wrote, "This is Copper Flat, the biggest ore deposit in the district." On the bottom postcard of the Veteran shaft mine, he wrote, "This is our head frame over the shaft. The main body of ore lies under the hollow in the center of the picture. This hollow has settled about 10 feet. The large barn in the center of the picture is now torn down. The front end settled so deep and the back end stayed up, that the floor was on such a slant that you had to crawl up on your hands and knees." Courtesy, Dunkle family.*

Although Seagrave was only ten to fifteen years older than Earl, he was an experienced underground mine manager. After gaining some experience in the deep mines of California's Mother Lode, Seagrave was one of about thirty shaft miners who were hired by John Hays Hammond to oversee the first deep mine in the Witwatersrand gold field in South Africa: the Robinson Deep.

Dunkle's appointment in a relatively junior position and Seagrave's experience in senior management suggest the importance of contacts in the worldwide yet small mining fraternity. After John Hammond left Africa, he became chief consultant to the Guggenheims. A few years later, Hammond was succeeded by Pope Yeatman, who visited the Alaska copper fields in 1909 and knew Seagrave from the African mines.[7] Yeatman probably recommended Seagrave, first for the Ely operation, then as mine manager to fine-tune the Guggenheims' just-acquired Beatson mine on Latouche Island in Alaska. When Seagrave left Ely for Alaska, Dunkle asked him to look for a mining position that would fit a young engineer. Seagrave remembered the request and sent for Dunkle, who landed in Cordova, Alaska, in early August 1910.

Earl Dunkle, who had grown up in the wooded plateaus of northwestern Pennsylvania, had not appreciated the arid Nevada landscape. He looked forward to Alaska with anticipation. As the steamship made its landfall on Hinchinbrook Island, bound for Cordova, the young miner was awestruck by the mountains that soared upward from the sea. Greenish black spruce forests transitioned upward to green alder thickets, then to emerald green tundra, with a few residual patches of shiny white snow. Dunkle never forgot his first Alaska landfall, the scene later painted for him by Sydney Laurence. Soon after his arrival in Cordova, Earl made a side trip to the end of the CR & NW Railroad track near Child's Glacier. It was only weeks after the completion of the Million Dollar Bridge. A few days later, Dunkle made another valuable acquaintance. He met Stephen Birch, the founder of the Kennecott enterprise.[8]

Dunkle's first Alaskan assignment, the Beatson mine on Latouche Island, was exciting. The unsophisticated twenty-three-year-old with open countenance was enthused about everything, whether it was working on the docks or riding a pitching skiff with boatman Billy Pay to Wallace's sawmill on the north end of the island. Earl Dunkle, however, did not neglect preparation for this beautiful but harsh environment. To become hardened to Alaska's cold waters, he regularly swam to the new loading pier at Latouche.[9]

In his job at the Beatson mine, assaying and surveying were not unfamiliar tasks for Dunkle, but the detailed geologic mapping of mines was new. Fortunately Dunkle had a graphic guide, the geologic map of the mine prepared two years earlier by L. A. Levensaler. Levensaler had been one of several young geologists who joined Horace Winchell when he assembled the first staff of mining geologists at Butte, Montana. Winchell's small crew devised a system of geologic mine mapping—the Butte System—which is still the basis for underground mine mapping. Dunkle picked up the Butte System from his study of Levensaler's geologic maps of the Beatson mine, a debt that he acknowledged in 1953.[10]

The Beatson lacked the glamour of the mines in the Wrangell Mountains. It had, however, an attraction of its own, and it was a mine of special importance to Birch and the Alaska Syndicate, both before and after production began in the Wrangell Mountains. Vessels of Alaska Steamship Company often arrived in Alaska fully laden but lacked freight for the return trip to Seattle. Concentrates from the Beatson mine furnished a back haul for "Alaska Steam" as well as feed for the smelting plant at Tacoma, Washington, before the rich Kennecott mines were in production. Beatson continued to fill a special role for the next two decades. Kennecott ore was almost too pure to smelt. The pyritic Beatson ore was an essential *flux* (see glossary) for smelting the nearly pure copper in Kennecott ores and concentrates.

When Dunkle arrived at Latouche Island, he found a bustling mining camp with some characters of dubious honesty. One of those characters was promoter Henry Derr Reynolds, whose most spectacular machinations occurred a few years before Dunkle's arrival. Silver-tongued Reynolds was so persuasive he forever clouded the reputation of Alaska's Governor Brady. He had sold Brady on a plan to construct the "Home" railway from Valdez to the copper mines. Reynolds was also the principal of Reynolds-Alaska Development Company.[11] Reynolds-Alaska developed the Horseshoe Bay sulfide deposit two miles south of and on the same mineralized trend as the Beatson, but without its rich concentration of copper. The scarcity of copper, however, did not prevent Reynolds from building a small town to develop the mine.

One of the island's attractive characters was Helen Van Campen. Petite, adventurous Helen arrived on Latouche Island at nearly the same time as Dunkle and remained a lifelong friend. Helen had just married Dunkle's immediate boss, mine superintendent Frank Rumsey Van Campen, one of several marriages for Helen. (In 1906 Helen had been Mrs. Green in New York City, where she wrote newspaper articles, short stories, and books and motored around the New Jersey Palisades with well-dressed friends in her Pope-Hartford Model G.) Helen continued to write for *McClures, Colliers, Everybodys,* and the *Saturday Evening Post* from a hideaway cabin near the mine. Her style, later compared to that of Ring Lardner and Frank McKinney ("Kin") Hubbard, was breezy, rich in dialogue, and poor in periods.[12]

Although Helen quickly adapted to Alaska, she was still accessible to her old friends, even from her remote outpost on Latouche Island. Samuel G. Blythe, novelist, newspaperman, and editor, put her accessibility to the ultimate test when he wrote Helen from St. Petersburg, Russia. The address was "Mrs. Helen Van Campen, A village, near a mine, on an island, Alaska, United States of America." The postal service connected with Helen after only one false try to the Treadwell mine on Douglas Island across from Juneau. Helen was totally fearless and a quick study of men's work on the docks and in the mines. She was also a suffragette and photographer, who captured Dunkle in many photographs on and around Latouche and in the Wrangell Mountains. She could even talk rough miners on Latouche into a group photograph and evidently make them enjoy it (fig. 8). Kennecott management, probably the austere and stuffy E. Tappen Stannard, later determined that Helen was too flamboyant and outspoken to be a good Kennecott wife.[13] Stannard's opinion, however, did not deter Dunkle's friendship with her.

Among the non-copper-mining families on Latouche were the Wallaces and the Kimballs. James Wallace had a sawmill on the north tip of the island. Beatson Copper Company bought most of Wallace's lumber to build an almost completely wooden camp (fig. 9). Dunkle later recalled that Wallace sold his lumber at eleven dollars per thousand board feet, a fraction of its price in the early 1950s.[14] The Kimballs—husband, wife, and young daughters, Vera and Decema—operated a retail store at the independent community of Phoenix, about a mile south of the mine. Occasionally Dee and her father boated to the Wallaces' sawmill, with Dee dressed in boylike foul-weather gear. James Wallace's son, William, was later pleasantly surprised to find that his playmate on those occasions was an attractive girl.[15]

In late 1911 W. H. Seagrave was promoted to general manager of all Kennecott mines in Alaska and moved to Kennicott, leaving Frank Van Campen in operational charge at the Beatson mine. At about the same time, Dunkle was offered a new job as field exploration engineer, or scout, for Alaska Development and Mineral Company of New York, which had an office in Cordova, Alaska. The company, directed by Stephen Birch, was the exploration arm of the Alaska Syndicate. Dunkle was the third man to hold the job, following L. A. Levensaler and J. F. Erdletts. Levensaler, field engineer in 1909 and 1910, quit after the 1910 season,

Figure 8. Copper miners on Latouche Island, ca. 1911. The photograph is from the Helen Van Campen collection. Well-known author Van Campen was married to Dunkle's immediate boss at the Beatson mine, Frank Rumsey Van Campen. Courtesy, APRD, Helen Van Campen collection, 74-27-436.

and the job was assigned to Erdletts. Erdletts did not meet Birch's expectations, and the position was vacant in December 1911.[16] Birch, undoubtedly with Seagrave's and Frank Van Campen's approval, filled it with Dunkle, whose apprenticeship was over. Between 1910 and 1930, Dunkle worked with several Guggenheim-related companies, as well as Alaska Development and Mineral Company and Beatson Copper Company. Later he was assigned to, or perhaps temporarily employed by, Bering River Coal Company, Alaska Steamship Company, the privately held Kennecott Mines Company, and its public successor, Kennecott Copper Corporation, for specific assignments. Newspapers of the time often said that Dunkle worked for the Guggenheims. The designation of Dunkle's employer could have been confusing. All of Dunkle's jobs had a common denominator, however—Stephen Birch—and Dunkle always stated that he worked for Birch.

Dunkle effectively held two jobs—exploration engineer and mine geologist—in the period from 1912 to 1915. When he was away from

Figure 9. The Beatson mine (top) *of Kennecott Copper Corporation, Latouche Island, and the Beatson mill* (bottom), *ca. 1917. Dunkle briefly was Beatson's mine superintendent in 1915 and 1916. Mine complex: Courtesy, Rogan Faith, private collection, Anchorage. Mill: Courtesy, APRD, Helen Van Campen collection, 74-27-56.*

Kennicott, Dunkle scouted for new mining opportunities for Birch's Alaska Development and Mineral Company. When he was at Kennicott, Dunkle was mainly a geologist. In later years, he appreciated the duality of his early career: "I found myself engaged in making the first detailed geologic study of the ore occurrences at Kennicott and was holding down the job of field exploration engineer for the Morgan-Guggenheim interests who owned the mine."[17]

Prospect examination, the major part of the scout's job, involves extensive travel. It is also challenging geologically. In most cases, exposures of ore are limited and may be largely covered by soil or younger rock. Moreover, when a district is first explored, the broader ore controls, which later are helpful in prospect evaluation, are unknown. Even small shows of mineral need to be developed (opened by mine workings) on the chance that they, like the Bonanza outcrop, might conceal large deposits. Dunkle mentioned these factors in his retrospective paper (1954) on the Copper River district: "[I]n a new country with no past mining experience to serve as a guide, a sure knowledge on which to base such interpretations can be gained only by the expenditure of time and money in development work. This was true in the Copper River country."[18] Because of the activities of the always restless prospector, Birch and his crew had many mineral deposits to examine. In the Wrangell Mountains, the area favorable for copper deposits had to contain the Nikolai Greenstone and the overlying Chitistone Limestone. As deduced earlier by the USGS, the area with these characteristics was large: at least fifty miles long and ten miles across. Within this area, there were hundreds of copper prospects, and Dunkle looked at most of them. The adjacent area to the south contained gold. Productive deposits of placer gold were developed at Dan, Chititu, and Young Creeks in the Nizina district. Somewhat more distant was the shoreline of Prince William Sound, where there were both gold and copper deposits (fig. 10). The rest of Alaska and the west coast of Canada also needed to be prospected and evaluated.[19]

In the spring of 1912, Dunkle made his first trip back to the Guggenheims' headquarters in New York City. His assignments on his return to Alaska were to visit the rapidly developing Willow Creek district in upper Cook Inlet and then continue southwestward down the Alaska Peninsula to examine prospects before returning to Kennicott. At that time, Willow Creek appeared to be the best lode gold district north of Juneau. Dunkle found an active camp in which production from lode mines had nearly eclipsed that of placer gold from the early discovered Grubstake Gulch and Willow Creek mines. Promoter William Martin

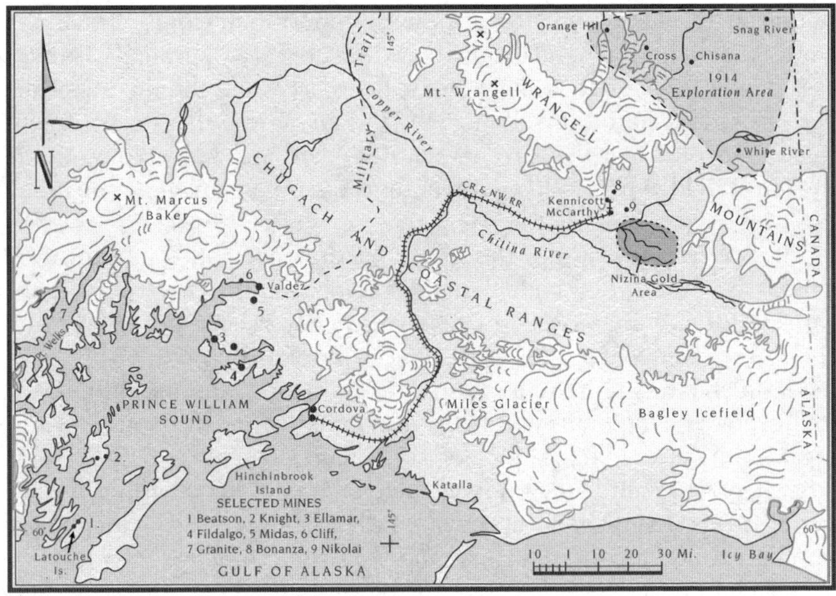

Figure 10. Prince William Sound and the Wrangell Mountains. Dunkle did much of his scouting and geologic work in this extremely rugged area between 1910 and 1916. Author's compilation.

was developing the Skyscraper mine, recently discovered by Robert Hatcher, and the Homestake vein. The most important discovery in 1912 was the Gold Bullion lode found by William Bartholf; it was rich, averaging about 1.7 ounces of gold per ton.[20] Although the district showed promise, access was difficult and the mines were small. Dunkle did not recommend any properties for acquisition by the Alaska Syndicate, but he kept the district under surveillance.

Dunkle's next stop was Chignik on the Alaska Peninsula, probably to look at the newly discovered copper and gold prospects in Prospect (Warner) and Mallard Duck Bays and the coal deposits at Chignik and Herendeen Bay. Dunkle left Chignik on the SS *Dora,* a sturdy little steamship with auxiliary sails nicknamed the "Bull Terrier of Alaska."[21]

The next leg of the trip furnished an unscheduled refresher course in volcanic geology. On 4 June 1912 the *Dora* began the return trip to Cordova, with stops scheduled at Karluk, Uyak, and Kodiak, all on Kodiak Island. As they sailed toward Kodiak, smoke, a deep haze, and occasional flashes of lightning appeared to come from an area to the north on the Alaska Peninsula. At Uyak on the southwestern coast of Kodiak

Island, residents reported hearing repeated explosions during the night. At 1 P.M. on 6 June, Dunkle was at the stern talking to Captain McMullen. Dunkle later reported, "Big black clouds suddenly appeared from the direction of Katmai that seemed to open and spread like a funnel, while a terrific belching noise was heard." A huge volcanic explosion had taken place. By 4 P.M. a great cloud of volcanic ash was overhead; by 7 P.M. it was pitch black, and the sky was raining a coarse pumice sand on the *Dora.* The wireless operator could not raise Kodiak because of static electricity in volcanic-dust-laden clouds. Since it would be impossible to make a landfall, the *Dora* set full speed on an open-water course for Seldovia on the mainland. The sky began to lighten about 10 P.M. In the meantime, the *Dora* had accumulated from six to as much as eighteen inches of coarse pumice. At times the vessel gave shelter to thousands of exhausted birds caught in the volcanic storm. The sky was black again for nearly three hours on the morning of the seventh and again in the afternoon, when the vessel was near Barren Island. After finally landing at Cordova, Dunkle gave the first authentic information about the eruption at Katmai. Based on the size of the volcanic cloud and the amount of pumice received by the ship, Dunkle estimated that some 70 billion tons of material were cast off in the eruption.[22]

Perhaps Dunkle should have anticipated some untoward happening when he boarded the *Dora* at Chignik. The little vessel was already legendary to those in the peninsular ports of call, but it was also usually late. *Dora*'s reputation suggested that it had been caught on every reef north of Seattle, and once it had been blown off course by one thousand miles. For Dunkle the clincher should have been an earthquake at Sanak Island in the Shumagins that reportedly was caused by the *Dora* actually arriving on time.[23]

Most of the prospects appraised by Dunkle between 1912 and 1916 were small or poor deposits that could be rejected immediately. Two that showed promise were the Midas copper prospect in Solomon Gulch about five miles south of Valdez and the Cliff gold prospect on the shoreline a few miles west of Valdez. The Midas mine, with two distinct lodes, was discovered by H. E. ("Red") Ellis and Mr. and Mrs. H. G. Debney. Ellis discovered the first lode and staked it as the All-American in 1901. Mary Debney then discovered and staked the nearby and more promising Midas lode. By the summer of 1912, early claimants of the mine had driven four hundred feet of *adits* (see glossary) into the Midas lode, and the property was under the control of George Baldwin through Midas Copper Company. Surveyor-attorney Baldwin had been the first general manager of Stephen Birch's development project in the Wrangell

Mountains. After Baldwin was replaced by R. F. McClellan, Birch continued to look after his former employee's welfare. When Baldwin submitted the Midas property, Birch agreed to its development. Birch's Alaska Development and Mineral Company leased the Midas claims from Baldwin in late September 1912, and development began immediately under Dunkle's direction.[24] Stephen Birch committed to an expenditure of at least fifty thousand dollars, enough money to more than double the previous work if machinery could be landed on the property. The latter was not a trivial feat.

To drive the new mine workings, Dunkle proposed to use air-driven rock drills powered by a distillate (a crude gasoline)-driven compressor. The compressor weighed several tons, and each hundred-gallon barrel of distillate weighed nine hundred pounds. Contractor Jim Wilson approached the job with a team of twenty-five Indian ponies that he called *cayuses,* one mule, sleds, and scrapers. After gathering the distillate, machines, and explosives on the beach, Wilson broke down the large machines as much as possible. Except for the crankshaft-flywheel assembly that weighed about two and a half tons, the other components of the compressor weighed about one ton apiece. Each barrel of distillate was a sled load for one of the ponies. To haul the heavy units, the contractor borrowed heavy freight sleds from Fort Liscum at the foot of the trail. The steep first mile of the trail was actually the easiest section because Wilson could push the snow off the trail with his scrapers. On the flats above the steep trail, the sleds wallowed in deep snow. From there to the mine, the teamster attached six-by-six-inch timbers about fifteen feet long to the heavy parts of the compressor. He also placed four or five men at the end of the outriggers to right the loads when the sleds departed from the track. As difficult as winter freighting was, it was better than summer conditions. In early May 1913, there was still ten feet of snow on the ground, but it covered an almost impenetrable thicket of alders fifteen to twenty feet tall (fig. 11).[25]

Dunkle's crew of twenty-five men drove more than one thousand feet of workings at the Midas mine during the winter of 1912–1913. The ore was there, although the ore zone was much narrower than at the Beatson mine. Using the price of fourteen cents per pound for copper, Dunkle calculated the operation as marginal for Birch, who therefore returned the property to Baldwin's Midas Copper Company in June 1913. Baldwin immediately sold the mine to Granby Consolidated Mining, Smelting and Power Company Ltd., a Canadian outfit with a tidewater smelter at Port Anyox, British Columbia. Dunkle's cost and profit projections, based upon the fourteen-cent copper price, had not allowed for World War I,

Figure 11. Freighting to the Midas mine in May 1913 with "Commodore Nelson and the good ship Helen*." The freight included mining equipment, explosives, and distillate (gasoline). Each horse pulled either a sled or one nine-hundred-pound barrel of gasoline. Courtesy, Dunkle family.*

when copper prices more than doubled and made the Midas a small but profitable mine for Granby. Baldwin made a commission of thirty-five thousand dollars on the transaction.[26]

Prospecting for copper on Prince William Sound declined between 1907 and 1914, when, because of wartime demand, copper activity soared again. A search for gold deposits more than picked up the slack in the intervening years. Some activity was induced by the success of the low-grade gold-quartz deposits in the Juneau district in Alaska's panhandle. The early success of the Alaska-Gastineau prospect at Juneau, briefly the world's largest gold mine, prompted prospectors to look for similar swarms of quartz veins along Prince William Sound. A prospector, now known only as Stokes, found three large quartz-vein systems in Pigot Bay near Whittier, Alaska, but Dunkle could not confirm Stokes's optimistic assays and rejected the prospects. Prospectors had more luck with simple high-grade quartz veins, but only a few of the hundreds of prospects proved to be worth developing. One of the better showings was the Granite vein, and Dunkle was at the Granite within days of its discovery.[27] By far the best was the Cliff vein near Valdez.

The small but rich Cliff gold mine, another discovery of "Red" or "Copper River Red" Ellis, made a good profit for its backers, to the chagrin of Dunkle's friend Henry Watkins. Red, an excellent woodsman and prospector, was strong and broad shouldered, with long curly red hair. He was a loner whom many people thought odd. Watkins and Ellis, however, had been friends since about 1900. Ellis showed Watkins twenty-five or thirty sacks of ore from the Cliff vein. All contained visible gold scattered through quartz, evidence of an extremely rich vein. Ellis proposed that Watkins put up three thousand dollars for a half interest in the property. The men would put in a one-stamp quartz mill, which was adequate for a small high-grade operation, and operate the mine themselves. Watkins knew the ore looked good, but he asked former Kennecott manager George Baldwin for his opinion. Baldwin told him, "Everybody knows Red is crazy," so Watkins backed out of the deal. A few years later, Baldwin—with Jim Lathrop, promoter B. F. Millard, and two or three others—bought into Ellis's claims, opened the Cliff mine, and made a $750,000 profit. Both Dunkle and Watkins thought the transaction was typical of Baldwin, but they did not bear him any hard feelings. Dunkle's later comment regarding the profit made at the Cliff mine reveals that he did not have a bad case of gold fever and that he understood the value of other resources. Dunkle noted that, although the Cliff made a one-time profit of $750,000, the cannery at Valdez made the same profit every year, which was not considered remarkable.[28]

Dunkle's first area of responsibility, as scout for Alaska Development and Mineral Company, covered the west coast of North America north of Seattle. His second, as mine geologist at Kennicott, included only a few square miles but was just as challenging. In 1912 there were few mine geologists in the United States, let alone in Alaska. Before about 1900, mine geology, if done at all, was part of the mine engineer's task. Most practical mining men thought geology was a waste of time because it got in the way of mining, but this attitude was due to change. In 1900 Anaconda Company established the first independent geologic department, and geologists soon proved their worth. One provision of the federal mining law, the extralateral right of a dipping vein, favored geologic competence over sole engineering skill. The extralateral provision gave the owner of a dipping vein the right to mine the vein through the sidelines of a mining claim, often under an adjacent rival claimant's ground. This opportunity meant that the claim had to be oriented correctly with its long direction parallel to the apex or lode line of the vein. In the early days of the mining law, many mining engineers successfully litigated extralateral rights, but they had to be better-than-average geologists to do so.[29]

One of Dunkle's first assignments at Kennicott was directly related to extralateral rights. After twelve years of prospecting in the area, it seemed clear that the most important ore bodies were in the Chitistone Limestone. Moreover, most of the ore in the rich Bonanza deposit was within three or four hundred feet of the base of the Chitistone Formation. In a general sense, an ore zone was parallel to the basal contact of the formation; thus, many of the earliest claims staked in the district followed the contact between the Chitistone and Nikolai Formations. But how was the ore deposited within the Chitistone? One young geologist, Ocha Potter, who worked for Kennecott's tiny rival Houghton-Alaska-Exploration Company, slipped into the Bonanza mine in the middle of the night to get a firsthand view of the ore body. Potter concluded that the early theory was incorrect: "I decided the ore body ran across the mountain range and not parallel to a limestone-greenstone contact as described in all official literature on the subject. If I were right, then the Kennecott claims had not been properly staked and government mining regulations would limit their mining to their side lines only, a few hundred feet down." Potter and Houghton-Alaska-Exploration followed up his surreptitious trip into the mine with staking. Some of their claims were oriented at right angles to Kennecott's earliest claim locations; if Potter was correct, they could have rights to some of Birch's ore.[30] Birch asked Dunkle for an independent opinion on the orientation of the claims. Dunkle quickly affirmed Potter's views. Most of the ore was in nearly vertical fissure veins with a northeast strike; in other words, nearly at right angles to the formational contact and the earlier staked claims. Dunkle recommended that Kennecott acquire Potter's claims and stake new claims in the Chitistone Limestone to cover the northeast fissures.

Dunkle had much more to learn about the geology of the ore bodies. Copper minerals were scattered throughout the upper part of the Nikolai Greenstone and were concentrated at the contact between the Nikolai and Chitistone Formations. The base of the Bonanza ore body, however, was nearly a hundred feet above the base of the Chitistone Formation. The lowermost one hundred feet of the Chitistone Formation was generally so barren of copper that it was called "the unfavorable lime." By chemical analysis, Dunkle found that the so-called unfavorable lime was, in fact, a nearly pure limestone—calcium carbonate—and that the overlying ore-bearing part of the Chitistone Formation was a dolomite that was much richer in magnesium carbonate.[31] More importantly, Dunkle also determined that the contact between limestone and dolomite, the base of the Bonanza ore body, was a fault that was almost parallel to the layers of the formation. In places the fault was a tight slip plane; elsewhere it was marked by a wider zone of broken rock. In his 1954 paper on the Copper River district, Dunkle described the geologic relationships this way:

> No copper was deposited in the lower limestone but as soon as the base of the upper lime was reached copper ore began to deposit and continued to replace the limestone up along the fissure to variable heights. For the most part, therefore, the greatest concentration of ore occurred at the very base of the upper lime. Deposition was aided at that point by the presence of a thick parting between the two beds of limestone. In this parting the lime was sheared and softened to a small extent, thereby giving the copper-bearing solutions a chance to spread out at this horizon. This parting was known at the mine as the Bedding Plane. It was continuous throughout the mine and was always found at the base of the ore or close to it.[32]

The ore bodies terminated upward in a zone of calcite crystals. Dunkle continued with his description of a vertical section parallel to the ore body, a section that showed "a continuous orebody, greatly elongated, with its top irregular and having the calcite zone as a sort of cockscomb along the top of it, the bottom following the Bedding Plane in a relatively straight line at a dip of twenty-two degrees, parallel to the greenstone and about ninety feet above it."[33]

Locating an ore body at Kennicott was an exciting event. Dunkle was there when the Jumbo was discovered, the second of three major ore bodies to be found in the area. On the surface, a northeast-trending fissure was nearly barren of minerals, so it was not staked when the first claims were located along the basal Chitistone contact. Exploration of the Jumbo began by driving an opening parallel to, but some fifty feet away from, the barren fissure. At intervals crosscuts were driven over to the fissure. Three hundred feet from the outcrop, a crosscut disclosed a one-foot vein of chalcocite in the fissure. Miners then cut back toward the surface on the newly found vein. At one round (about five feet) on their new heading, the vein had widened to three feet. At the next round, blasting disclosed a full face of solid chalcocite, as black as coal. Miners continued *drifting* (following the vein) back to the bedding-plane fault. They were in solid chalcocite for 120 feet. When they reached the fault and crosscut to the edge of the ore, they found that the vein had widened to eighty feet. In map view, the ore body was a pear-shaped mass of solid chalcocite. On the surface, however, there had been no indication of ore except for a barren, nearly vertical fissure that crossed a strong bedding-plane fault.[34] Based on the price of copper and the shape and size of the ore body indicated by exploration, Dunkle calculated the value of the new ore body as at least $26 million, almost exactly what it cost Kennecott to build a railway, construct a mill town, and establish the Bonanza mine.

The discovery of the Jumbo ore body took place shortly before the formation of Kennecott Copper Corporation as a public company in 1915. Although the ore produced from the Bonanza ore body was fabulously rich, the Alaska Syndicate had not made a profit on its immense investment to develop the mine and its infrastructure. Because of the high copper price during World War I and the enthusiasm generated by the discovery of the rich Jumbo ore body, stock offered in 1915 for the newly formed Kennecott Copper Corporation was fully subscribed. The Alaska Syndicate recovered its original capital essentially overnight.[35]

Almost immediately after the Jumbo discovery, Dunkle headed for New York to brief Stephen Birch about the progress at Kennicott. The two men were having lunch at the company headquarters when Daniel (possibly Murry) Guggenheim walked in. Birch introduced the young geologist to Guggenheim, who asked about news from the north. Thinking that Guggenheim would have heard of the Jumbo discovery, Dunkle referred to it in an offhand way. It was apparent that Guggenheim had not heard of it. Birch said, “Tell him about it, Dunk.” Dunkle proceeded to do so, but Guggenheim apparently was not too impressed. Dunkle’s later reaction was one of awe. The Guggenheim mines were so numerous and so rich that a discovery that would rank among the greatest single ore bodies ever found did not rate much attention at the company headquarters.[36]

The exploration technique used in the discovery of the Jumbo—driving parallel to a suspected ore body and systematically crosscutting to the hoped-for ore—was typical and effective at the well-managed Kennecott mines. From 1912 to 1915, Dunkle and his Yale classmate DeWitt Smith produced the first geologic maps of the Kennecott deposits and helped their mining leaders—Seagrave and general mine foreman Melvin Heckey—to find new ore.[37] Dunkle's geologic contributions were significant. He appears to have been the first to notice the connection between dolomite and the ore at Kennicott, as well as the importance of the bedding-plane faults that localized the base of the major ore bodies. He also noted that the almost certain source of the copper was the underlying Nikolai Greenstone. He proposed that the energy that drove mineralization was provided by an intrusive porphyry south of the major ore bodies. Dunkle's immediate successors, Alan M. Bateman and D. H. McLaughlin, furthered knowledge of the ore controls at the Kennecott mines. Bateman and McLaughlin found that the steep fissures that controlled the strike of the ore bodies were in and parallel to minor *synclinal* (down fold) structures. The minor structures were cross folds to a major up fold (*anticline;* see glossary), a structure that was partly eroded. The geologists proposed that the formations would tend to pull apart in the synclinal cross folds and thus be more susceptible to ore-bearing fluids.[38]

It is difficult to imagine a more energetic individual than Dunkle in his role as scout and geologist, especially as he added other chores to an already busy schedule. Dunkle worked with Archie Hancock on a survey of the tramline that connected the Bonanza mine to the mill at Kennicott and completed most of the survey himself. Before the arrival of metallurgist E. Tappen Stannard at Kennicott, Dunkle worked with Earl Hinckley on devising new methods to leach the copper carbonate ore, which had been largely lost in processing. At times general manager W. H. Seagrave called on Dunkle for extra service because of Dunkle's athleticism and stamina. On one occasion, air pressure for ventilation at the Bonanza mine was lost, creating a life-threatening situation. All of the miners on the shift, except for Con Kelly, who had remained on the surface, passed out because of a lack of oxygen. To assess the problem, Seagrave sent Dunkle up a three-thousand-foot climb of more than three miles length, an ascent that he made in fifty minutes.[39] Fortunately ventilation was soon reestablished, and there were no fatalities from the incident. (To assess Dunkle's feat, consider the following: Dunkle's contemporaries, William Shedwick and Henry Watkins, both excellent hikers, took an exhausting three hours to follow the same route downhill.)

Some of Dunkle's seemingly boundless energy was spent in hiking. In those days, walks of one hundred or more miles along well-used Alaskan trails were routine. Some of Dunkle's hikes, however, were solo and on routes that were only then being pioneered. An example is the hike that Dunkle made to assess the newly discovered Chisana district in 1914. In the preceding year, prospectors found rich placer gold deposits at a site about seventy-five miles north of Kennicott, causing a stampede to the discovery, which was soon named Chisana. Kennicott's miners left in droves for the new district, and for a while Kennicott looked like a ghost town.[40] Most of the miners returned when it became evident that mineable gold deposits were confined to two short creeks, but the new district still had to be examined.

Dunkle arrived in Chisana on 3 July 1914. He spent most of the next two months looking at Chisana and nearby camps. One group of prospects, thirty air miles east of Chisana near the snout of the Nabesna Glacier, was nearly sixty miles away by trail, through Cooper Pass and down Cooper Creek to the Nabesna River. At the site, Dunkle found disseminated copper at Orange Hill on the Nabesna and extensive mineralization at a greenstone-marble contact at nearby Iron and Nikonda Creeks. The best showing, however, contained only about 5 percent copper, which was not rich enough for development at such a remote locality. At an even more distant site at Cross Creek, fifteen rugged mountain-and-glacier miles southwest of the Nabesna, massive lead- and zinc-rich skarn contained no redeeming values in gold and silver to justify development. Copper and gold were also reported from Snag Creek, fifteen miles northeast of Chisana across a seven-thousand-foot-high pass. Dunkle hiked over the pass with R. F. McClellan, the tough old Minnesota timber man who constructed much of early Kennicott. For once Dunkle could not pull his own weight. He strained a muscle in his back, and McClellan had to take two packs over the pass. The back problem disappeared as rapidly as it had begun. Dunkle was never sure if McClellan had accepted his explanation or had merely thought that it was the malingering of a young college man.[41]

Strength and technical competence were key to Dunkle's work as scout and mine geologist, two rather lonely jobs. Some scouts were probably more sociable. A few seem to have done most of their business in bars through buying drinks for prospectors, but that was not Dunkle's style. His work at two projects during his early Kennecott days suggests that Dunkle already had considerable leadership skills to add to his technical ability; it also suggests that Dunkle approved of Stephen Birch's management style and probably tried to emulate it throughout his own career. Dunkle's first opportunity to display management ability was at the Midas project during the 1913 season. His second opportunity was at the Beatson mine in late 1915 and early 1916. The Midas job was simpler, but its less complex demands still covered a wide range of activities: acquiring supplies and equipment, delivering them to the site, hiring contractors, and foremost, leading a twenty-five man team consisting of miners, two cooks, probably an engineering assistant, a mechanic, and laborer support. At the close of the project, Dunkle evaluated the results and made an economic forecast. He decided that the project was not viable at a fourteen-cent-per-pound copper price, but the work was done competently and probably within Birch's fifty-thousand-dollar budget.

Sometime afterward, as Dunkle completed the evaluation of the Golden Zone prospect in the fall of 1915, W. H. Seagrave offered him the mine superintendent's job at the Beatson mine after Frank Van Campen left. Beatson was a large mine; it was, in fact, larger in tonnage than the Wrangell mines, although it was leaner in grade. Mining was mainly by *glory-holing* (see glossary). The mine employed hundreds, and the town included a store, hospital, wireless station, and company housing for miners and staff—all of which Dunkle had to manage. Every week or so, ships arrived at Beatson to load ore and copper-rich concentrates for the smelter at Tacoma, Washington.

Dunkle was replaced at Beatson in 1916 by the favorite of the new general manager, E. Tappen Stannard. By that time Dunkle had established sufficient rapport with his miners that they offered to strike so that he could keep his job.[42] At Beatson and at his later projects at the Lucky Shot and Golden Zone mines, Dunkle was supported by his miners.

Dunkle had especially good relationships with his shift bosses, foremen, and skilled electricians, smiths, and mechanics. In the Kennecott years from 1912 to 1915, Dunkle thought very highly of two men: Melvin Heckey, the general mine foreman, and Carl Engstrom, a versatile smith and mechanic. Heckey played a critical role in the discovery of the Jumbo ore body. He also found an extension to the Bonanza ore body when he, in direct contravention of Birch's orders, drove additional workings after the ore was lost at a fault. Heckey had a "nose for ore" that Dunkle thought reflected systematic observations and logical deductions: deductions that Heckey probably could not articulate but could implement.[43] Engstrom was the best among the skilled mechanics and smiths at the mine. Dunkle admired Stephen Birch for selecting good men and delegating operating authority to them. In the 1930s Dunkle built up his own team and gave his men room to make appropriate decisions.

In 1912, when Dunkle assumed the scout's job, he had advanced considerably in the Kennecott organization, had made several good friends, and had worked very hard at jobs that he liked. He still lived a rather lonely existence, however, a situation that was soon corrected. On one of his trips to the Cordova office of Alaska Development and Mineral Company in the summer of 1912, Earl Dunkle met Florence Hull, who soon became his fiancée. Florence had just been hired as a secretary by Stephen Birch.[44]

Although Dunkle met Florence Hull in Alaska, in many respects she was the girl next door. Florence's father, Linn Walker Hull, like Earl's father, was a lawyer of exceptional skill and ethical standards.[45] The Dunkle and Hull grandparents lived on homesteads in the country south of Lake Erie. Florence was born in Sandusky, Ohio, about seventy miles west of Earl's boyhood home. Many family values were shared. The Dunkles and Hulls were Protestant, Republican, and committed to education. The Hulls in particular were committed to Oberlin College. Florence, her older sister, Marguerite, and her younger sister, Emily, had attended Oberlin College, as had her father and several aunts, uncles, and cousins. Earlier Hulls may not have shared the fierce Methodist piety of Dunkle's grandmother, but there is little doubt that the Oberlin Hulls would have endorsed Caroline Boyer Dunkle's regard for Abraham Lincoln. Oberlin's record as an antislavery institution is unmatched.[46] Judge Linn Walker Hull would certainly have agreed with Caroline. Hull, as a young attorney in private practice, and his partner E. B. King successfully defended an indigent black man from the charge of first-degree murder.[47]

Florence graduated from Oberlin College in 1908 with a bachelor's degree in education. Like Earl, she was a class officer and was active in college life. She was honored as her class historian and as president of the women's literary society, Phi Alpha Phi. Fellow students thought well of her: "[She] is exceedingly bright, [concerning] both books and conversation. She can lead literary society with perfect ease and ability. 'Tis hoped she will marry a governor so that her qualities may have ample scope. But it's against her principles to take advantage of leap-year privileges. She thinks for herself."[48]

After graduation Florence continued to assert her independence and to see quite a bit of the United States. Florence taught in Ohio, Minnesota, and Iowa and at a government school in Fort Huachuca, Arizona, a location close to the bonanza silver camp of Tombstone.

Both Florence and her sister Marguerite taught in the silver camp at Wallace, Idaho, where Florence may have met her first Dunkles.[49] Earl's uncle Peter and a small clan of Dunkle immigrants from Pennsylvania operated Dunkle Gardens, an enterprise that supplied produce to much of the Silver Valley of northern Idaho.

Dunkle and Florence had more in common than just family background. Both were adventurous and independent. They read together, and each had a love for and ability in music. Earl played the flute throughout his life. Florence was a serious pianist, who had spent her first years at Oberlin in the Conservatory of Music; she later taught her older son, John, how to play classical pieces before he was ten years old.

When Earl met Florence Hull in 1912, he was twenty-five years old and had an almost certain career with Birch and the Guggenheims. Florence was two years older and had traveled extensively, but she was willing to make a major career change. She abandoned a secure but probably impecunious career as a teacher for a position as the secretary of one of the most powerful men in Alaska—Stephen Birch—a position that must have led to a rather irregular courtship. Both Florence and Earl had "floating" offices. Florence's location depended on Birch's location; Dunkle's location depended on his particular assignment. The young couple, however, must have met often at either Cordova or Kennicott. By the summer of 1914, the couple set a date near Christmas for a wedding in Oberlin, Ohio. Florence left Kennicott first, while Earl was still at Chisana. Earl was in Seattle on 9 December completing his report on Chisana and other prospects, but he arrived in time for his wedding on 22 December. The wedding, noted in an Oberlin newspaper as the "Culmination of Alaska Romance," added extra sparkle to the holidays. With the exception of Dunkle and his father, the wedding was very much an Oberlin affair. J. A. Barber, Florence's uncle and an Oberlin graduate, gave the bride away. The couple was married by Rev. John Henry Hull, another Oberlin graduate, and the wedding was held at the home of Florence's sister Marguerite Hull Badger, an Oberlin alumna. Twenty-two family members and friends attended the wedding.[50]

The young couple left on the evening of 23 December for a honeymoon in the eastern United States. The Oberlin newspaper noted that they would be at home in Cordova, Alaska, in February.

4

Life and Geologic Challenges at Kennicott

Aristocratic society has its upstairs and downstairs. Mining society has its uphill and downhill. The uphill stratification of residency begins with lowly surveyors and young engineers and moves upward through mine and mill superintendents to the general manager, the level commanding the best view. Miners live downhill on muddy or dusty streets. Storekeepers, shift bosses, skilled mechanics, and foremen occupy a transitory middle ground, sometimes up, sometimes down.[1] Earl and Florence Dunkle, who were already on the lower rungs of the uphill ladder, continued to move up.

The idealized housing scheme outlined above was physically impossible at the Kennecott mines near McCarthy, Alaska. Because of the location of the ore, the miners had to live uphill in bunkhouses clinging to the mountains, more than three thousand feet above the mill town of Kennicott. The mine manager, superintendents, engineers (including Dunkle), office personnel, and skilled technicians lived downhill in Kennicott. The mines—at first only the Bonanza but later the Jumbo, Mother Lode, and smaller Erie and Glacier mines—were linked with Kennicott by miles of aerial tramway.

The village of McCarthy, four miles south of Kennicott along the CR & NW Railroad, had no official ties to Kennicott but was an integral part of its life. It was the headquarters for the extended mining region. Copper prospectors came in from Clear Creek, Glacier Creek, and Peavine Bar to promote their wares. Some met Dunkle in McCarthy to show him maps and ore specimens in order to entice him to visit their prospects and perhaps option them for the Alaska Syndicate. Gold miners, including Stephen Birch's brother Howard and others from Dan, Chititu, and Young Creeks in the nearby Nizina gold placer district, arrived to exchange information and pick up freight, supplies, and new hires. McCarthy was the "downhill" destination of miners on their rare days

Figure 12. Kennicott mill, ca. 1920. The dark building in the foreground is the general manager's office as remodeled in 1912; the white building is a staff house built in 1910. The view is to the north-northwest toward Donoho Peak; moraine-covered Kennicott Glacier is to the left. Courtesy, Alaska State Library, Juneau, Early Prints of Alaska, PCA 01-3115.

off. It had two bars and a few extra girls. Even there, however, one bar was "uphill," reserved by style and ownership, if not decree, for managers and mine promoters.[2]

Kennicott was a company town managed as a well-planned headquarters for production of copper ore. The largest and most important building in the town was the mill (fig. 12). Dunkle's office was in the staff house. There was a bunkhouse for bachelors, but by 1915 several cottages had been built for married staff. At least in terms of the local topography of Kennicott, the general manager's office was at the highest elevation. The CR & NW Railroad and a parallel wagon trail connected the mill, power plant, storehouses, and sawmill with McCarthy and the more distant world.

Kennecott's miners lived in their uphill villages at the mines. They had warm quarters (which occasionally shifted as the permafrost thawed), good food, billiards, music, and ultimately a motion-picture theater. The well-staffed and well-equipped hospital was downhill at Kennicott, even though it was needed most often by the miners.[3]

As far as Kennecott Copper Corporation was concerned, the total purpose of the enterprise was to produce copper and ship it to Tacoma as efficiently and quickly as possible. As long as miners were willing to work seven days a week for a wage that Kennecott deemed reasonable, the company was happy to look after them in a paternalistic fashion. The no-nonsense style of Kennecott management fit Dunkle. His mind was challenged by geologic and engineering problems that took at least a twelve-hour day to solve. In later years Dunkle could not break away from a schedule that called for breakfast at 6:30 A.M. seven days a week. It was not an Alaskan city schedule, however. A person could stand in the middle of the street in Seward or Anchorage at 9 A.M. or later on Sunday morning and not see a soul, whereas Kennicott was in full operation.[4]

Work was assuredly what most people, whether miners or professionals, did at Kennicott during the first days of the camp. The schedule during the construction years under R. F. McClellan, almost four years before Dunkle's arrival, was twelve hours per day seven days a week without break. As did most Americans, miners celebrated the Fourth of July and Christmas, but all other days were workdays. As construction ceased and the mine began to produce in 1911, W. H. Seagrave instituted a change: His small crew of professionals could have Sunday afternoons off.[5]

Earl Dunkle and Florence Hull were not at Kennicott until 1912, but in 1909, as the mines were preparing to open, their friends Henry Watkins, L. A. Levensaler, and Larry Bitner had begun a tradition of evening socializing that lasted for years. The practice was extended to the broader Kennicott community in the next generation after the Dunkles. Inger Ricci (née Jensen), born in Kennicott in 1918, remembers tennis, organized ice skating, Saturday night dances, and elaborate festivities on Christmas and New Year's Eve as the high points of her youthful life. Inger still cannot conceive of a happier place to grow up. Her father was carpenter-foreman, one of the skilled middle class who lived in Kennicott.[6]

In the earliest days, the camp was almost entirely a bachelor affair. Birch's stenographer, Larry Bitner, could not take shorthand, but he could write fast and was able to keep up with Birch's dictation. At first storekeeper Henry Watkins handled most of the other office chores, including payroll. Women began to arrive as the first homes were completed. Mrs. Shrock accompanied her husband, Elisha B. Shrock, M.D., who was hired as company physician in 1909. (Shrock's nurses had a high turnover rate, rarely staying single more than one season.) Archie Hancock, mechanical maintenance engineer, and his wife, Ada Jane, arrived in the summer of 1911, as did Larry Bitner with a red-haired bride, who helped him handle an expanded job of payroll and accounting. Larry's earlier stenographic job was taken by Florence Hull.[7]

In late 1911 W. H. Seagrave replaced A. B. Emery, who had only served a few months, as general mine manager. Seagrave brought his very young daughter, Ellabelle, and her new stepmother, Molly, to the mine. Seagrave retained the engineers, including William John Shedwick Jr. ("Shed"), whom Emery had hired for the project. Shed, a near contemporary of Dunkle, planned to alter his marital status. In July 1912 Shedwick wired his fiancée, Margaret Ely: "Can you meet me in Seattle October 13th to get married?" Margaret ("Peggy") wired back in the affirmative. A honeymoon reception planned for the newlyweds at Kennicott reflects the innocent horseplay that young engineers indulged in at the remote mine. On their return to Kennicott, a merciful Archie Hancock met the train a few hundred yards from the station and spirited the young couple away. Hancock told them, "Hurry out and let's sneak between the line of cars to my house as all the boys in camp are waiting to haze you. They plan to lock Mrs. Shedwick in the Club House and you in the ice house." Other hazing of new engineering employees was almost as innocent, such as the so-called badger fight. A badger attached to a long rope was placed in a box, open side down. A new employee, whom Shedwick called a *chichauker (cheechako),* was given the opposite end of the rope. Then, according to Shedwick, "a dog was brought in and placed in front of the box ready to pounce on the badger. Considerable betting took place as to which would win the fight. . . . When everyone was well keyed up, the guest was asked to pull the badger out. Unfortunately there was no badger on the end of the rope—only a potty."[8]

On the uphill side of Kennicott, "solo" and poker were initially the most popular card games. Seagrave, because he was familiar with the bad effects of large gambling debts on small crews in remote locations, set limits at poker for his professional crew but not for the miners. The engineers had to play penny ante with a ten-cent limit, a limit that Dunkle also adopted at the Beatson mine in late 1915 during his short tenure as mine superintendent.[9] There was no way, however, to police the miners' bunkhouses for gambling infractions, where the sky was the limit at poker and craps. In later years contract bridge became the game of choice at uphill Kennicott, leaving poker for the miners.

Two annual events—the washout of the railway bridge above Chitina and the Fourth of July baseball game at McCarthy—were soon celebrated as traditions. Kennecott miners bet on both. For the baseball game, each team brought in the best pitchers possible, all the way from Seattle if necessary. A few of the miners, including master blacksmith Carl

Engstrom, made better use of their time producing useful and artistic works from colorful blue-green copper ore. Engstrom, earlier a mechanic for Buffalo Bill Cody, could fix anything from a watch to a locomotive.[10]

Music was a nightly affair, possibly in the miners' bunkhouses but definitely in the lair of geologists, engineers, and office staff. Mrs. Havemeyer, Stephen Birch's original patron, furnished a Victrola, and Birch had Watkins pick up about fifty records whenever Henry made a run to Seattle. The musical taste of office and professional staff extended at least into the light classics. Dunkle, in an address book that dates back to the early 1920s, listed twenty-three of the old favorite records that he found at Kennicott when he returned briefly in 1924. In addition to *Souvenir,* played by violinist Maude Powell, there were several waltzes, *Humoresque* by Kreisler, a Paderewski version of *Minuet in G,* and two arias featuring Caruso. A few pieces, including "Happy Heine," "Too Much Mustard," and "The Spaniard That Blighted My Life," were on the much lighter side.[11] The engineers and office staff also made their own entertainment. Larry Bitner's specialty was poetic parody; in the early years, L. A. Levensaler acted as a nightly master of ceremonies.

Soon after the arrival of the first automobiles at Kennicott (and in Alaska), Sunday picnics or trips a few miles long proved immensely popular. Dunkle and Watkins remembered that, for several years, just the opportunity to go for a drive anywhere was exhilarating. When he was stuck in Ketchikan for a prospect examination, Dunkle might have been invited for a drive to Ward Cove. In Cordova Florence Hull may have been invited on a Sunday drive to Eyak Lake. The CR & NW Railroad scheduled occasional sightseeing trips, and Kennecott employees, such as Shed and Dunkle, could travel on the rails in gasoline-powered rigs or on tricycle speeders (fig. 13).

Fishing and small-game hunting were popular recreation for almost everyone. Throughout his life, Dunkle thoroughly enjoyed bird hunting and fishing. But like many other nearly full-time residents of rural Alaska, he saw little reason for big-game trophy hunting. It was one of the few areas in which Dunkle did not care to emulate Stephen Birch. His boss was a serious big-game hunter. In fact, Birch refused to stop one hunt for such a minor event as a world war. In late July 1914, Birch, Henry Watkins, two Johnson brothers, and a Japanese cook left Kennicott on an extended hunt for mountain sheep. Before dawn on 1 August, a horseman galloped into camp with the news that war had broken out in Europe: World War I. Birch said, "Boys, we'll finish the hunt, then I'll go back to New York and make a million dollars." The men hunted for three more weeks;

Figure 13. Dunkle on CR & NW Railroad speeder, ca. 1924. Courtesy, APRD, Helen Van Campen collection, 74-27-525.

then Birch left Kennicott for New York to make more than a million dollars when Kennecott Copper became a public corporation in 1915.[12] Shares offered at twenty-five dollars sold for fifty-seven dollars within six months.

Birch shared his good fortune with his Alaskan engineers, including Shedwick, who had moved to Tacoma to have direct control over Kennecott Copper's transactions with the smelter. Birch kept one hundred shares of the new company for Mr. and Mrs. Shedwick and helped them finance the purchase. After repaying Birch, the Shedwicks had a $5,200 nest egg, which allowed them to buy their first home and an automobile, "a four cylinder Hupmobile with wire wheels and there was only one more in Tacoma like it." Dunkle received similar consideration and held the shares for many years. In 1937 part of Dunkle's income was five hundred dollars from Kennecott dividends.[13]

The management crew in Dunkle's years at Kennicott (1912–1915) was small (fig. 14). Even in later years, the senior mine-mill staff probably never exceeded fifteen men. As the mines were developed, however, more miners were needed, and their number eventually exceeded five hundred.[14] Some were experienced miners from the Rocky Mountain west, but many were immigrants, and turnover was fairly high. Because of telegraph communication to Seattle, the CR & NW Railroad,

Figure 14. Engineers and mine staff at Kennicott, ca. 1915. Left to right: *Frank Rumsey Van Campen, Becker, Dick Watkins (?), W. H. Seagrave, Stephen Birch, Dunkle, "Doc," and Shuey. (The full names of Becker, Watkins, "Doc," and Shuey are unknown.) Courtesy, Dunkle family. The photograph is identical to APRD, Helen Van Campen collection, 74-27-427N.*

and Alaska Steamship Company, a fairly constant supply of new hands was available at Kennicott. For many new arrivals, Kennicott was a shock. After the miners drew their equipment and supplies, they went uphill to the bunkhouses and mines via the aerial tramway. (At least in later years, mine visitors and new employees signed a release absolving the company from any accident before their tram ride.) Entry to the tramline was from a station at the top level of the mill. Each man stepped into a bucket and was on his way, in some places hanging three hundred feet above the ground.

Some recruits did not like the aerial prospect at all. The alternative of drawing no pay and returning to Seattle penniless was not pleasant either; most miners hid their fears and took the first ride. The second and subsequent rides were easier. Engineers also rode the trams and occasionally had problems. In the wintertime, the trips to the mines and the mines themselves were cold, and men dressed in maximum layers. Shedwick almost froze to death when the tram stopped for several hours. On one trip upward to do some mapping at the Bonanza mine, Dunkle dressed in too many layers to be flexible. He missed the bucket but finally pulled himself into it, a difficult chore, especially because he could see the tram boss laughing hysterically at his plight.[15]

The tramline was only one risk of many. Especially in the early years, danger showed little respect for anyone, whether miner or executive. The miners were exposed to rockfalls and premature or faulty blasting. Everyone was at risk from avalanches and frigid Alaskan waters. At the end of the first construction season, most of the men returned to Seattle. Barker, who constructed the first sawmill, remained in camp with James Montgomery, an adventuresome and well-to-do lad from Yale. The next spring, in an effort to hasten the construction of the Bonanza tramway, the first workers shoveled out a linear trench in the snow along the tram alignment. Lacking support, snow avalanched into the cut. It caught and killed Barker, Montgomery, and Carl Engstrom's partner, Gray.[16]

Rivers presented the worst hazards. Before the railway was completed, most of the heavy freighting was done in the wintertime with horse-drawn sleds that followed frozen rivers. Teams and men frequently broke through the ice. Experienced freighters could usually improvise ways to save men and equipment, but sometimes rescue and salvage were impossible. Henry Watkins fell through the ice on the Chitina River and was in the water, almost at head height, for fifteen minutes. Watkins gave instructions for his last rites and provided his father's address in Virginia, but he was finally saved when a sled was lowered into the water. Henry was too numb to catch a rope.[17] Watkins, Dunkle, and all the early travelers learned as much as they could about Alaskan waters, including various methods for crossing white-water rivers in the summer, but all river travel was extremely risky.

The Kennicott years were some of Dunkle's best. Friendships initiated at Yale with DeWitt Smith and Dave Irwin were renewed, and Dunkle worked with men whom he deeply respected: Birch, Seagrave, Melvin Heckey, and Engstrom. The years were especially productive geologically. In later years, Dunkle thought more like a mining engineer. At Kennicott, he had the opportunity to work as a geologist. He was as concerned with the processes that had localized the ore tens of millions of years before as he was with the methods involved in delivering the ore to the concentrator. Intellectual curiosity, as well as the need for more ore at the mill, drove Dunkle to work on the geology of the rich copper mineral, chalcocite.

The Kennecott copper deposits may be geologically compared and contrasted with those in the Keweenaw Peninsula of Michigan. Mines in the Keweenaw Peninsula exploited rich deposits of native copper—almost pure copper metal—in basaltic flows and interlayered volcanic conglomerates. In the Kennecott mines of Alaska, fabulously rich veins of chalcocite were deposited within a limestone formation that lay above a thick series of basalt flows. Other examples of these two deposit types exist, but none have been as important economically as the Keweenaw and Kennecott deposits. These two deposits have other similarities and contrasts. In 1850 the Keweenaw was the first bonanza copper camp in the United States. Of those discovered during the early twentieth century, the bonanza deposit at Kennicott was the last.[18] In the Keweenaw mines, Cornishmen—the Cousin Jacks—taught Americans how to mine underground lodes. Seventy years later the American pupils were the teachers, exporting mining knowledge to the rest of the world.

Alaskans chauvinistically tend to overestimate the importance of their Kennecott mines in the copper world. Kennecott's porphyry copper mine in Bingham Canyon, Utah, produces in two or three years the same amount that Kennecott's Alaskan mines produced in twenty-seven years (1.2 billion pounds of copper). Moreover, the Bingham Canyon mine was in production before the mines in Kennicott, Alaska. The copper deposits at Chuquicamata, El Teniente, and Escondido in Chile dwarf the deposits in Alaska, as do those of the African copper belt. At that particular moment in time, however, Kennecott's deposits were remarkable. The Bonanza and the later discovered Jumbo and Mother Lode mines were incredibly rich. In the earliest days, train-car loads from Kennicott consistently averaged about 70 percent copper. Although the Kennecott mines produced less copper than Butte, Montana, on a tonnage basis, they produced much more per man. In 1916 Kennecott Copper Corporation in Alaska produced ten million pounds of copper per month. At Butte Anaconda Company produced three times as much. Butte's production, however, came from thirty working shafts and fifteen thousand men. Kennecott's production came from low-tonnage mines with not more than 550 men. Kennecott easily won the horsepower battle. Butte's mines drew thousands in horsepower to pull ore and ground water up from deep shafts. Kennecott's Alaskan mines developed almost no groundwater, and as soon as the ore hit the tramlines, gravity took it all the way to the bottom of the mill. Even the tramlines were developed as electric generators, producing about one-quarter of the power needed for the mines. From its remote site, Kennecott could land copper in New York at slightly less than five cents per pound. For a time other copper miners were worried that the Alaskan Kennecott mines would kill copper prices worldwide.[19]

In the first years of the new century, Alaskan copper flashed as a meteor across the night sky. When Clarence Warner and Tarantula Jack

Smith looked uphill from the mouth of Kennicott Glacier in 1900, their eyes were caught by a bright green patch high on the mountain south of the glacier. The green color was caused by a copper carbonate mineral, malachite, rather than their first guess, a patch of grass. When the prospectors arrived at the green outcrop and broke the rock, it was evident that the malachite was a mere patina on a bluish steel-gray mineral, chalcocite. It was an incredible deposit, but if Professor L. C. Graton of Harvard University was right, there was a problem in the chalcocite itself.[20] Graton said that the chalcocite was *secondary,* that is, it formed by the replacement of a primary mineral with a lower copper concentration. The implication was that, as depth was gained in the lodes, chalcocite would give way to bornite or perhaps chalcopyrite, minerals that contain much less copper than chalcocite. The grade of the ore would drop precipitously. At a mine in Montana or Nevada served by a transcontinental railway and a nearby captive smelter, such a change in mineralogy would be serious but survivable. In remote Alaska, where the Alaska Syndicate had invested more than $25 million in development and infrastructure before production began, it could be fatal.

At Kennicott, young mining geologist Earl Dunkle was in the middle of this controversy. Dunkle thought that chalcocite was *primary* (deposited by ascending *hypogene* fluids) and that the mines would remain profitable at depth. His views, however, were counter, not only to Birch's consultant L. C. Graton but to the views held by most of the copper industry. At the mines at Bingham, Utah, and Ely, Nevada, chalcocite was secondary. At depth, below the chalcocite blankets, the copper porphyry ore was low-grade chalcopyrite, which was then too lean to mine. At Butte, Montana, Anaconda's chief geologist, Horace V. Winchell, also believed that chalcocite was secondary. The issue was important enough to Winchell that Anaconda Company engaged in industrial espionage to find out more about the copper deposits in Alaska. In 1903 and 1904, Winchell mounted an expedition to Alaska (fig. 15). In addition to Winchell, the expedition included Anaconda counsel Capt. 'D Gay Stivers, a geological consultant furnished by John Hays Hammond named J. D. Audley Smith, and young L. A. Levensaler, who worked as an assistant to both Winchell and Smith. The expedition was guided by experienced Alaskan wrangler and explorer George Hazelet. Winchell visited many copper deposits, including the Beatson on Latouche Island, but Stephen Birch made certain that Winchell had only a cursory view of the Bonanza mine in the Wrangell Mountains. Perhaps if Winchell had studied the Bonanza, he would have revised his opinion. With his limited knowledge of the ore at Kennicott and his own prejudice favoring the super-

Figure 15. Horace V. Winchell, Anaconda Company, on a scouting expedition to Alaska in 1903. Anaconda wanted to know more about Alaska copper. Courtesy, APRD, Julia Sweeney collection, album 3, 97-139-41N.

gene (secondary) origin of chalcocite, Winchell concluded that the Bonanza's rich ore was limited to shallow depths and of little importance to Anaconda.[21]

The controversy over chalcocite has long been resolved, but during the first thirty years of the twentieth century, the issue was of wide scientific and economic interest. In his summary article, "Recent Progress in Studies of Supergene Enrichment," included in the classic first Lindgren volume, *Ore Deposits of the Western States,* W. H. Emmons wrote this: "In the earlier studies of secondary enrichment it was commonly believed that essentially all chalcocite was supergene. . . . Economically the issue was so important that it continued to occupy the attention of most geologists who were working with copper ores in the field and in laboratories."[22] Because of its critical importance to copper mining, copper producers funded the Secondary Enrichment Investigation, directed by Professor Graton. The resolution of the controversy, attained at Kennicott in about 1920, was the result of the first systematic collaboration between theoretical geochemists (scientists at Carnegie Geophysical Institute, Washington, D.C.) and mineralogists and economic geologists in academia and industry.

At Kennicott from about 1913 to 1915, the clash was between the empirical mine geology of Dunkle and the theoretical studies in laboratories in Washington, D.C. and Harvard University. Dunkle's opinion on chalcocite was based on observations in the mines. He did not believe that the mineralogy changed with depth. Graton's views were based on both the newly developed technique of ore microscopy under reflected light and the rapidly emerging science of theoretical geochemistry. In papers published in 1915 and 1916, geochemists affiliated with the Carnegie laboratory proposed that chalcocite crystallized in two different systems, depending on the temperature of formation.[23] High-temperature chalcocite crystallized in the cubic (isometric) system. Low-temperature chalcocite formed in the orthorhombic system. Each type had characteristic crystal forms. Even when high-temperature chalcocite cooled through an inversion temperature and its atomic structure changed, it retained remnant triangular patterns characteristic of its cubic origin. (The triangular pattern represents another cubic form, the octahedron, which is produced when the corners of a cube are cut.) The triangular patterns could be enhanced by etching and were then visible on polished surfaces under the reflecting microscope. Bornite, however, is also a cubic mineral. If it had been replaced by chalcocite (as the woody tissue of a tree is replaced by silica to become petrified wood), similar cubic patterns could be inherited. Graton believed that the characteristic triangular structure was inherited from bornite and that Kennecott chalcocite was secondary; his word carried much more weight than a young field geologist-engineer. Dunkle's successors in the study of the mines at Kennicott—Alan Bateman and D. H. McLaughlin—used a combination of field and laboratory data to conclude that Kennecott chalcocite was hypogene (primary). The men found the characteristic triangular patterns of high-temperature chalcocite in their own laboratory investigations (fig. 16). They also thought that the discovery of natural octahedral crystals at the Bonanza mine was especially convincing evidence of hypogene chalcocite.[24]

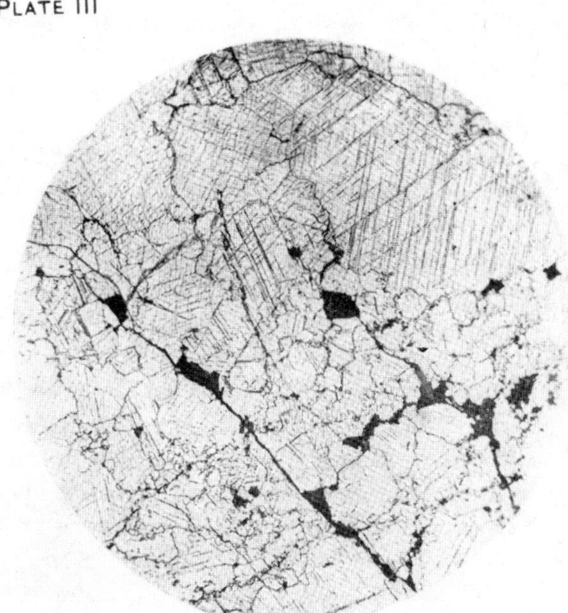

Figure 16. Photomicrograph of polished and etched chalcocite with triangular pattern. Geologists Alan Bateman and D. H. McLaughlin believed that the triangular patterns supported the hypogene (ascending) origin of the chalcocite. From plate 3, Bateman and McLaughlin, "Geology of the Ore Deposits."

Modern scientists have found further complications. The high-temperature chalcocite of Bateman and McLaughlin at Kennicott is now known to be a separate mineral species called *djurleite.* There is a similar temperature implication as well because djurleite also develops a characteristic etch pattern as it cools: the same structure that the earlier scientists believed was diagnostic of high-temperature chalcocite.[25]

The Kennecott enterprise was a huge gamble in almost every respect. The Bonanza ore body was the erosional remnant of a deposit. In 1913 the Jumbo deposit was still unknown. All of the deposits were remote; access would involve crossing huge rivers and glaciers. And, of course, there was the chalcocite problem. It is probably fortunate that Stephen Birch, the Guggenheims, and J. P. Morgan Jr. did not know much about copper mineralogy during the decision days. They did know there were outcrops of nearly pure chalcocite. They won the gamble.

Both practice and theory ultimately showed that the Kennecott chalcocite was primary, and new ore was discovered at the Jumbo and Mother Lode mines.

Dunkle later used his acquired knowledge of copper mineralogy to assess other copper deposits for Kennecott, including some as distant as Africa. He also used it in his fight to acquire the Mother Lode, the continuation of the Bonanza deposit at Kennicott.

Stephen Birch and his backers began the Wrangell Mountain copper enterprise on the basis of the Bonanza deposit, which was incredibly rich but of uncertain size. To maintain the enterprise, more rich ore had to be found, the closer to the Bonanza the better. In 1906 Clarence Warner and Jack Smith, who discovered the Bonanza, found a new deposit four thousand feet northeast of the original discovery. They called it the Mother Lode and located their first claim as the Marvelous. Two questions about the deposit soon emerged: Was it important, and who owned it? The Mother Lode was at least hundreds of feet above the base of the Chitistone Formation, whereas the Bonanza was only one hundred feet above the base. Furthermore the Mother Lode appeared to be much smaller than the Bonanza, and its ownership was questionable. Despite these concerns, Dunkle believed that the deposit had great potential and that Kennecott should acquire it and correct any deficiencies in its title.

The chalcocite controversy, albeit one of scientific interpretation, hurt Kennecott because the uncertainty temporarily dampened the enthusiasm that Stephen Birch had for the area. The Mother Lode issue was equally important, but it had human instead of scientific roots. Warner and Smith stayed on in the area and worked for Birch as caretakers during the early winter seasons before the mine opened. They were probably Birch's employees when they discovered the Mother Lode. Kennecott could have reasonably claimed at least partial ownership because of either the employer-employee relationship or the extralateral rights of the Bonanza deposit.[26] Warner, Smith, and a third man named Oscar Sales, however, staked the Mother Lode in their own names, and Kennecott did not contest the claims.

A strange and tragic event in the life of Oscar Sales complicated the title to the property. Sales broke through the frozen Chitina River and tumbled downstream under the ice. He escaped, but mentally he was never the same. Sales disappeared on a trip "outside." Warner and Smith arranged for Jim Godfrey to promote the Mother Lode, but the uncertainty over Sales's death clouded the mine's title and affected Godfrey's efforts to either develop or sell the mine.[27]

In 1912 the only ore body known with certainty at Kennicott was the original Bonanza. The Jumbo was a raw prospect, and an uncertain Mother Lode was controlled by others. Shortly afterward, Graton dropped his chalcocite bombshell: that the rich ore could be limited to the near surface. More than $25 million had already been invested in the mine and its infrastructure because Birch's backers believed that the Kennecott ore would be the high-grade chalcocite discovered by Warner and Smith. Tentatively accepting Graton's belief, Birch elected to pursue a conservative strategy: to stay with demonstrated ore in the Bonanza rather than acquire nearby speculative properties, such as the Mother Lode. That strategy would minimize losses if Graton's views on the limited extent of high-grade ore proved to be right. The Mother Lode, which could have been acquired cheaply, was thus left to Warner, Smith, the uncertain Sales estate, and Jim Godfrey to promote as best they could.[28]

Dunkle's own theories on the origin of the Kennecott ore were being developed at that time. By the fall of 1912, Dunkle had determined that the Bonanza ore had been deposited in dolomite rather than pure limestone. He theorized that it was important to determine the eastern limit of the dolomite in the Chitistone Formation. By then it was midwinter and too dangerous to explore the peaks east of the mine. He could, however, safely explore the Chitistone in the Mother Lode workings, nearly a mile east of the Bonanza, if he could gain access to the mine. As Dunkle later explained it, "The owners of the two mines were at loggerheads but the operating personnel were not, and I was able to get permission to take some rock samples from the Mother Lode workings." The rocks were dolomite. A few weeks after his visit to the Mother Lode, Dunkle determined that the base of the ore in the Bonanza was a subtle low-angle fault lying almost parallel to the bedding layers of the formation. Furthermore, the ore-bearing rocks were weakly broken, or *brecciated,* along other bedding zones. Again Dunkle thought of extending his observations in the safety of the Mother Lode mine, but by that time relations between the mine owners had deteriorated so much that Dunkle was denied permission for another underground visit. Later he resented the attitude of the Mother Lode crew, who sent him back to Kennicott on a safe but roundabout thirty-five-mile trek at twenty-five degrees below zero, without access to the mine and with nothing to eat since his breakfast in Kennicott.[29]

By projecting the base of the ore from the Bonanza to the Mother Lode, Dunkle estimated that the Mother Lode surface prospects were about eleven hundred feet higher in the Chitistone Formation than was the ore in the Bonanza mine. The ore at the Mother Lode was spotty, but

Dunkle reasoned that, if there was still ore hundreds of feet above the more favorable zone of brecciation at the base of the Bonanza, there could be a major ore body beneath the surface.

Dunkle later relates, "Early in the spring of 1913, I . . . found a very wide zone of brecciation, which included about half of our Azurite claim and most of the adjacent Marvelous Claim of the Mother Lode. . . . I was much impressed by the geologic evidence there and again tried to get Mr. Birch to take favorable action on the Mother Lode." (Dunkle made his first plea after he recognized the favorable dolomite strata at the Mother Lode.)[30] Birch was still in a conservative mode and, for a time, the Mother Lode was left to its claimants.

Although he took no action at the time, the question of the Mother Lode was definitely on Birch's mind for several years. In 1911 Guggenheim consultant H. A. Keller examined the Mother Lode for the Alaska Syndicate. In July 1912 Birch requested a copy of Keller's report. Carl Ulrich, Birch's young New York secretary, made sure that the report was on the way: "I sent you this report July 19th, but to make assurance doubly sure, I am enclosing another copy of this report to you at Kennecott."[31] In December 1913 Dunkle wrote Birch about patent proceedings at the Mother Lode, referring to the Oscar Sales story: "Replying to your letter of November 19th relative to 'Mother Lode' patent proceedings. This property has not been surveyed for patent. The 'Mother Lode' people wished to have the survey made but decided it would be useless to do so until after seven years had passed from the date of Oscar Sales' disappearance, which will be in just two years."[32] Birch replied promptly to Dunkle: "I do not know the source of your information but I wish you to make certain that there are no proceedings at this time, and the information which we have now is correct. You might also ascertain the status of Houghton-Alaska patent proceedings. They have claims surrounding the Mother Lode property, and in fact one right in between the Mother Lode's main claims."[33] Birch sent Dunkle a more detailed confidential letter the same day:

> One of the first things I wish you to do when you return to Alaska is to go into Kennecott and make a complete map showing all the Kennecott claims including the Jumbo, and the Mother Lode and Houghton-Alaska Companies claims. . . . On this same map I wish [you to include] the workings of both the Jumbo and the Bonanza; and where the Mother Lode people are working. Give courses and distances wherever you can. After you have made this map I would like one showing where the workings of the Bonanza mine would

> intercept the Mother Lode location if it continues on the same strike and dip as the present workings indicate, and at what depth below the Mother Lode workings. . . . Keep this information to yourself and only consult Mr. Seagrave as to what you are doing.
>
> I am enclosing herewith a copy of Keller's report showing a map of the claims. I think Levensaler furnished this map. If, during the course of your work, it becomes necessary for you to state why you want a map from either Levensaler or Adams, say that you are compiling a general map of the mining districts in Alaska (which as a matter of fact I think it well for you to have). I think that Levensaler could get a map direct from Potter if he asks for it. Potter has charge of the Houghton-Alaska property and he told me last summer that he overlaps some of our claims around the Jumbo. As you are probably aware Potter is not very friendly to our people, which of course includes myself, for that reason I am not asking for the map direct and it will not be necessary for you to ask for it if you can get it in some other way.
>
> This work is quite important and you can start on it at any time, the sooner the better.[34]

Problems at the Mother Lode were resolved by 1919. Thereafter the mine was operated by a joint venture called Mother Lode Coalition Mines Company. Kennecott was the operating partner. Dunkle did not see the Mother Lode from 1915 to 1924. In 1924 the Mother Lode ore body was more than one thousand feet long and as much as eighty feet wide, and it extended hundreds of feet above the base of the dolomite. A longitudinal section parallel to the deposit shows the main Bonanza–Mother Lode ore zone about one hundred feet above the contact with the Nikolai Formation. A second small ore body, the original Mother Lode on the Marvelous claim, is stacked above it: exactly the relationship that Dunkle inferred many years before. The 1351 and nearby *stopes* (see glossary) on the Mother Lode ore body have been described as the richest ever mined.[35] Kennecott made a joint venturer's profit from the Mother Lode but probably would have made even more if Dunkle's earlier recommendations had been followed.

5

Scouting for Birch

Earl Dunkle produced his best geologic work in the period from 1912 to 1915, when he was the part-time resident mine geologist at Kennicott. He did not, however, neglect his contemporaneous work as scout for Alaska Development and Mineral Company. During the same years, Dunkle visited and evaluated more than one hundred named prospects in a region that stretched from Idaho (one examination), through British Columbia (thirty-six examinations), to Alaska.[1] In 1912, Dunkle's first year as a scout, all of his examinations were in south-central Alaska near his headquarters at Cordova and Kennicott. In 1913 he expanded his range into southeastern Alaska and British Columbia. In 1914 his examinations included those made on a two-month-long sortie into the Chisana area. After Earl and Florence returned from their honeymoon in early 1915, they embarked for Dunkle's northernmost (to that date) incursion into Alaska: the Broad Pass area on the southern flank of the Alaska Range.

Dunkle's progression from coastal Alaska to the central Alaska Range paralleled that of earlier prospectors, who entered the region either from Cook Inlet or the Copper River. Commencing in Cook Inlet, prospectors found placer gold at Hope in 1894, Willow Creek in 1897, Yentna in 1905, and farthest inland, west of Broad Pass by about 1907. The prospectors who followed the Copper River found gold at Chistochina (Alaska Range) in 1899–1900 and then discovered the Nizina district in the Wrangell Mountains. They renewed their advance into the Alaska Range, where they made a major discovery at Valdez Creek in 1902.[2] The discovery resulted in a minor stampede in 1903.

Because of the remoteness of the Broad Pass region, word of discovery leaked out slowly. Even recording their claims (registering them officially with the district commissioner) required a 150-mile dogsled run or an equally long summertime poling-boat trip to Knik on the north shore of Cook Inlet. Nevertheless, a rush to Broad Pass began in about 1911 and lasted through 1915. Distance, perhaps, magnified the reality of the new area, described by some overly optimistic people as the "Rand of Alaska," after the Witwatersrand of South Africa, the world's greatest gold district.

The leaders in the rush to the new region were two brothers, Alonzo ("Lon") and Frank Wells, prospectors who came into the country from Idaho in time for the rush to Valdez Creek. The men were part North American Indian, exceptional woodsmen, and already experienced at the mining trade. Unlike many prospectors, who blindly followed the next stampede, the Wells brothers stayed in the central Alaska Range and Talkeetna Mountains to prospect. Soon after, they began to prospect near Broad Pass, where the brothers joined forces with another experienced prospector and woodsman, John Coffey. Coffey probably had already discovered placer gold in Bryn Mawr Creek, a north-flowing minor tributary of the West Fork of the Chulitna River. By 1909, but certainly by 1912, the Wells brothers and Coffey had zeroed in on the lode source of the Bryn Mawr placer. Veins rich in quartz and sulfides laced a low knob, Rusty Hill, at the head of Bryn Mawr Creek, where the three partners located the Golden Zone Nos. 1 to 3 claims on 12 September 1912. By 1 November 1914 there were 114 claims owned by more than twenty-five individuals and syndicates in the Broad Pass area. The main block of claims blanketed a northeast-aligned belt four claims wide (about half a mile) and eight miles long. The land play must have been noticed at Guggenheim headquarters. In the late summer of 1914, Stephen Birch urged Professor Ransome, a placer expert from California, and Tom Mack, an engineering assistant, who were examining the Valdez Creek placer mines, to grab some samples from Broad Pass on their return to Knik. At about the same time, Birch told Dunkle to try to get there from the south, but only after he had completed his scheduled examinations. Because of the lateness of the season and several unfinished assignments near Kennicott, Dunkle could not get to Broad Pass. However, he did meet Tom Mack at Knik, and the men compared notes on the new area. To be on the safe side if a major rush developed, Thomas P. Aitken, a mining man with close ties to the Guggenheims, optioned (acquired the right to purchase) what appeared to be the key claims at Broad Pass in the fall of 1914.[3]

Samples brought back by Mack and others from the new area showed a considerable range in gold value. Most were low grade, about a tenth of an ounce of gold per ton. Some, however, contained as much as an ounce of gold per ton. The samples, collected over a large area, were definitely of interest, but there was a complication. Both Dunkle and Mack suspected that at least some of the samples had been salted.[4]

The grade, or richness, of deposits of the more common metals, such as copper, zinc, or lead, can be estimated from the proportion of visible minerals. Visual estimates of the richness of gold ores, which are considered high grade at one ounce of gold per ton, are almost impossible. The samples must be assayed to determine grade. Unscrupulous prospectors or assayers can "salt" a sample by adding bits of gold to it so that the assay indicates gold where none is present in the uncontaminated sample. Two assay offices, one at Seward and one at Knik, processed samples from Cook Inlet and adjacent areas. Experienced prospectors John Coffey and the Wells brothers brought in legitimate gold-bearing samples from the Broad Pass area. Less experienced men trailing them collected ordinary rocks and turned them in for assay. The assayers guessed that the prospectors were in a promising gold area, and they were determined to profit in two ways. First, each assay cost five dollars; prospectors were more likely to turn in additional samples if assays indicated that gold was present. Second, the assayers grubstaked some would-be miners, then "validated" their claims with fictitious assays; the prospectors shared the proceeds with the assayers when the claims were sold.

Dunkle based his suspicion of salting on the appearance of the samples. An acquaintance of Dunkle had collected samples from the new district that looked like barren dolomite but supposedly contained about eight dollars in gold (about 0.4 ounce per ton). To check the validity of the assays, Dunkle hovered over assayer Harry Ellsworth of Seward, but the samples still ran about five dollars per ton. Dunkle and Tom Mack virtually proved salting when they rented the Knik assay office, carefully prepared their samples, and found much lower values. Nevertheless, the scheme could not be completely discredited until the company assayer at Latouche reran Dunkle's samples. Most were barren. Assays of five closely guarded samples selected from visibly mineralized claims ranged from three to eighteen dollars in gold per ton (about 0.15 to 0.9 ounce), a significant amount that justified a field visit and more extensive sampling.[5] It was too late in the dark winter season to make the trip. The examination was rescheduled for the earliest time when daylight and snow conditions permitted and when Earl and Florence had returned from their honeymoon.

The Dunkles pushed their schedule. Originally planning to arrive in Alaska in February, they returned in late January, bypassing Cordova for Seward. The *Cordova Daily Alaskan* noted the revised destination and also that "their many friends in this section will be glad to give them a hearty welcome home." At the time, the best way to reach the upper Cook Inlet country was by boat into Seward, then up the track of the on-again, off-again Alaska Northern Railroad to Kern at the head of Turnagain Arm.[6] From Kern several gas boats, including the *Traveler* and the *Chase,* took passengers and freight to Ship Creek, Susitna Station, or Knik, depending on their final destinations. Water travel was in the daytime and correlated with the tides, which were among the highest in the world. Above the trading posts at Knik and Susitna Station, late winter was the best season for travel. Dog teams still had snow to traverse, and mosquitoes, apparently much fiercer than those of the present day, had not emerged from their watery homes.

The upper inlet was still icebound when the Dunkles arrived in Seward. They left Seward by dogsled in mid-February 1915 and crossed the Chugach Mountains above Indian River. Beyond the range, at the mouth of Eagle River, they found ice thick enough to cross upper Cook Inlet to Knik. The trip took ten days, but there were roadhouses along the way, so it may have been a passable continuation of their honeymoon.[7]

The examination of the Wellses' claims and other Broad Pass properties, which today could be conducted in a few days from Anchorage via helicopter or truck, then required an expedition approach. Four tons of equipment and supplies for the expedition had been sent to Knik toward the end of the previous shipping season. When Earl checked the inventory, however, he found that some critical items were missing, requiring a trip back to Seward.[8]

Florence had originally planned to wait for Earl in Knik, but the ice broke exceptionally early that year, so dogsled travel and schedules were uncertain. The best option seemed to be for a young Guggenheim engineer named Cooper to start north by dogsled with the supplies and equipment at hand and for both Earl and Florence to return to Seward, where Dunkle could pick up the rest of the outfit. Florence could catch the boat for Cordova and wait for Earl's return in more comfort. The trip to Seward by boat sounded both quick and easy, but Henry Watkins later told Dunkle in what was an understatement, "You had a nasty trip that time."[9]

The trip started easily enough. Earl and Florence mushed a few miles down to Goose Bay, where they joined Capt. Sam S. Cramer and an engineer as the only passengers on the gas boat *Traveler.* Leaving Goose Bay about 1 March, they had an uneventful first day's run across Cook Inlet. On the second day, however, as they approached a constriction in the inlet between the East and West Forelands, the *Traveler* became icebound. The vessel's progress over the next few days was largely determined by the tide. Each day Cramer advanced the boat a mile or so, but it took five days to get out of the icy trap and turn the corner toward Seward. Then, what should have been an easy one-day motor up Resurrection Bay, turned into a five-day nightmare. At one point Cramer, who was comfortable only in the more protected waters of Cook Inlet, panicked and tried to pull into a shallow bight, thinking that it was a safe anchorage. Dunkle managed to talk him out of that move. On another occasion, Dunkle convinced Cramer to hold a course across tidal rips, which ran as high as thirty-five feet in every direction, to safe glassy water in Port Dick. The supposedly easy trip of two or three days from Goose Bay to Seward took ten. During the last five days, Florence had been able to keep only "a couple of oranges down."[10]

On their arrival at Seward, Florence boarded a Cordova-bound vessel, whereas Dunkle departed for the north after grabbing the necessary equipment. On 22 March he was back in Knik on a catch-up schedule. He left Knik with experienced musher Jack Cronin within a day or so of his return. Traveling light, Dunkle and Cronin almost caught up with the Guggenheim engineer, Cooper. They had skimped on groceries and were almost out of food when they reached Ohio Creek, where they ran into the Nugget Kid. Nugget, who was returning to the Bull River at the north edge of the Broad Pass country, was also running low, but he did have a big block of chocolate, which he shared with them until they accessed their own supplies at Chulitna. Dunkle thought that the mushers whom he had hired for Cooper—John Reichert, George Vause, and the Harper brothers, Alfred and Charles—had done an exceptional job, and he told them so. When Jack Cronin returned to Knik, he said that W. E. Dunkle of the Guggenheims was an "A+ traveler and an indefatigable worker."[11]

At the West Fork of the Chulitna River, below the Golden Zone prospect and near the Riverside claim, Dunkle found two embryonic towns, both on the north side of the river. Wellsville was on one side of south-flowing Colorado Creek where it entered the river. Colorado City was on the opposite bank of the creek. Another settlement, which consisted of a store and roadhouse, opened at Indian River, thirty miles east of Broad Pass at the navigational terminus of the Susitna River. The trading post was "easily reached by pack train" during the summer months.[12]

Construction of two larger and more permanent projects began during the Broad Pass stampede. In late April 1915, the *Mariposa, Admiral Evans,* and *Latouche,* all vessels of Alaska Steamship Company, were docked at Ship Creek on the east side of Cook Inlet, discharging men and equipment for the construction of a town and railway. Ship Creek was hurriedly and officially renamed Anchorage. Lt. Frederick Mears, one of three designated commissioners for the Alaska Railroad, arrived at Ship Creek on the *Mariposa.* The two other commissioners, William Edes and Thomas Riggs, were expected to arrive within the week on the *Alameda.* Mears wasted no time in laying the first track from the beach to the site for shops and warehouses. Within the next few weeks, all of the hundreds of tents and tent buildings erected since March were torn down and replaced by more permanent buildings for the railroad and town.[13] As in a gold rush camp, the railroad-project community was quite diverse, with a cast of characters from Nome, Skagway, Dawson, Ruby, Iditarod, and other "places too numerous to mention." Alaskan pioneers, eager to check out a new venture, numbered in the hundreds.[14] They included one old Dunkle friend, Helen Van Campen, who was photographing the developing city, and a new Dunkle friend, Arthur A. Shonbeck. Shonbeck, an early stampeder to Nome and Iditarod, established a retail, wholesale, and outfitting business that prospered as Anchorage grew.

Dunkle's first examination of one property at Broad Pass—the Golden Zone—was sufficiently encouraging that he sent back for additional supplies and powder to begin tunneling on the Golden Zone No. 3 claim. By late June his crew had driven 150 feet of underground workings and planned to go another 125 feet. As reported in *The Cook Inlet Pioneer,* "The Guggenheims . . . are confining development to this one claim, because it will, in large measure, test others in the same mineralized zone."[15] Most other claimants were passively waiting for Dunkle's results, but the ambitious Wells brothers and John Coffey drove workings on and sampled the Riverside and Northern Lights claims. Dunkle or his men sampled at least eight other properties in the area. To avoid any chance of salting, Dunkle assayed all the samples himself.

During his examination, Dunkle kept his sled dogs working. Dogs that had pulled sleds to Colorado Creek were given the job of sample carriers; each dog packed a twenty-pound load back to camp each evening. Dunkle had a big malamute husky named Whitey. One day during a river crossing, Whitey proved his worth. The West Fork of the Chulitna River is a broad glacial river with multiple channels. Unless the Chulitna is in flood stage, it is usually possible to locate a foot-crossing through one channel at a time. On one particular day, sampling had been done on the north side of the West Fork, opposite the camp. On return to camp, all the dogs except Whitey stayed with their handlers and walked up and down the channels until they found a place shallow enough to ford. Whitey, however, apparently shared his master's straightforward nature and made a beeline for camp. The dog disappeared underwater with his heavy load of samples. Although Dunkle thought that Whitey had surely drowned, he went downstream to pick up the samples. Before long the big husky emerged nonchalantly from a channel, shook himself off, and continued his trip to camp. Evidently Whitey had just walked downstream on the channel bottom in the direction of the current until the river shallowed.[16]

The first official account of the Chulitna area (Broad Pass) was published by S. R. Capps of the USGS in 1919. This report was based upon a quick but remarkable wartime reconnaissance of the region. Capps was handicapped, as were most of the USGS topographers who worked on wartime projects. With his small horseback crew, Capps produced a reconnaissance map that showed the beltlike nature of deposits in the northern part of the Chulitna region. Most of the deposits were confined to a series of red beds that Capps mapped as tuff, limestone, shale, and lava flows. USGS paleontologist T. W. Stanton dated the series as Triassic.[17]

In the same year, Capps also completed reconnaissance surveys of the Willow Creek district and the western Talkeetna Mountains, essentially following Dunkle's trips of 1912 and 1915. Both Capps and Dunkle, however, were following prospectors, notably the Wells brothers and John Coffey. The prospectors evidently enjoyed local fame. They appeared prominently in the *Knik News* and its successor, the *Cook Inlet Pioneer,* and explorer Belmore Browne thought the men "represented the finest type of Alaskan frontiersmen." Browne recognized John Coffey's primacy in crossing the Alaska Range, and Dunkle was impressed by the rough terrain covered by the men. Dunkle photographed Frank Wells on a ledge that he admitted was tough: "Wells, the owner of the claims I went to examine. Crossing a particularly bad spot on 'O.K.' Mountain. I do not care for this kind of work when there is snow" (fig. 17).[18] The location of O.K. Mountain is not given, but it looks more like Iron Creek in the Talkeetna Mountains than Broad Pass.

Dunkle pulled out of Broad Pass in 1915, possibly as early as 20 July. Others in the camp, including experienced miners such as William Springer, thought that the Guggenheims had not done sufficient testing. At the time, however, Dunkle's analysis was correct. Some of the rock encountered in the tunnel in Rusty Hill assayed more than a quarter ounce of gold per ton. The average grade over a distance of about two hundred feet, however, was slightly less than a tenth of an ounce, which was insufficient to work in the remote district. Dunkle recognized the Broad Pass area as a large and potentially valuable district that would

Figure 17. Frank Wells on O.K. Mountain, possibly near Iron Creek in the Talkeetna Mountains, pre-1930. Wells disappeared without a trace in the 1930s. Frank, his brother Alonzo, and John Coffey discovered the Golden Zone and many other deposits. Dunkle wrote, "I do not care for this kind of work when there is snow." Courtesy, Dunkle family.

require rail access to be economical.[19] The Golden Zone prospect and the Broad Pass area were placed on the shelf, perhaps to be looked at again after the completion of the Alaska Railroad. Railroad construction began about 1 July 1915 from the port at the mouth of Ship Creek, the newly named city of Anchorage.

The Dunkles also had other priorities. Earl and Florence Dunkle's first son, John Hull Dunkle, was born on 14 October 1915 in Seattle. At about the same time, Dunkle accepted the mine superintendent's position at the Beatson mine on Latouche Island, where he had begun his Alaskan career five years before. The young family moved to Beatson. Unfortunately events largely beyond Dunkle's control made Beatson a short-term home.

Ammonia, often used as a common household cleanser, indirectly caused Dunkle's departure from Beatson, as well as the departure of general manager W. H. Seagrave and general mine foreman Melvin Heckey from the Kennecott organization in 1916. The three men were close personally as well as professionally. Seagrave had hired Dunkle at Ely, Nevada; had brought him to Alaska as a junior engineer at Beatson; had approved his appointment as exploration engineer; and in late 1915, had appointed Dunkle mine superintendent at Beatson. Heckey and Dunkle worked together on operations and ore searches at the Bonanza and Jumbo mines. Dunkle thought that Heckey was the best practical miner he had ever met, and Dunkle was Heckey's favorite among a group of college-trained engineers at Kennicott. The man ultimately responsible for the loss of these three men was a brilliant metallurgist, E. Tappen Stannard. Stannard knew how to recover copper with ammonia, but he was never particularly good with men whom he thought beneath him.

The incident stemmed from two roots: one financial, the other technical. During the first two years of Kennecott's operation in the Wrangell Mountains, the price of copper ranged from sixteen to seventeen cents per pound, which was sufficient to make a very good operating profit because Kennecott could land copper in New York at a total cost of five or six cents per pound. In 1914 the average price slipped a bit as World War I cut off supplies to a copper-consuming Germany, thus temporarily increasing supply and decreasing cost elsewhere. Copper stocks were dropping rapidly, however, and it was clear that, if World War I continued, the copper demand and price would soar. Stannard convinced Stephen Birch that if he (Stannard) assumed management and hired his own staff, the Bonanza and Jumbo would produce more copper than they did under Seagrave's management.[20]

The technical root of the controversy lies in the character and metallurgical reçovery of the Kennecott ore. The first shipments consisted of nearly pure chalcocite mined from massive ore bodies. The ore was mined selectively and shipped to the smelter at Tacoma, Washington, without further concentration; hence, it was called *direct smelting ore.* Lower-grade but mineable quantities of chalcocite and other copper minerals were disseminated in dolomite at the edge of the massive ore bodies and, in total, contained a large amount of copper. The disseminated deposits were rich enough to mine and then mill or concentrate; hence, they were called *mill or concentrate ore.* The first mill, or concentrator, at Kennicott was small and simple. It processed about three hundred tons of ore per day in simple gravity circuits (consisting of jigs and shaking tables) that concentrated the dense copper sulfides and rejected the less dense waste rock or gangue.[21]

As the masses of solid chalcocite were depleted, it became more important to mill the lower-grade material. In addition, parts of the ore body, from top to bottom, contained copper carbonate minerals—brilliant blue azurite and green malachite—that were hardly being recovered. The solution to Kennecott's problems in recovering the copper carbonate minerals and disseminated sulfide ores and in maximizing the mill efficiency was E. Tappen Stannard, who graduated from Sheffield Scientific School in 1905. Stannard's reputation started at Federal Smelting in Idaho and grew when he solved milling problems at El Teniente

(Braden mine) in Chile.[22] By this time mill deficiencies were evident at Kennicott. Stephen Birch asked Seagrave if he wanted Stannard, and Seagrave asked Dunkle for his opinion. Birch was prepared to pay Stannard five hundred dollars per month, which then was not a bad wage for a mill man on his third job. Dunkle told Seagrave, "He'd be a very poor mill man if he couldn't save that much in this mill."[23]

Stannard arrived at Kennicott in late 1913 or early 1914. By 1915 he had solved essentially all of the mill recovery problems, both at Kennicott and at the Beatson mine on Latouche Island. Stannard first upgraded the gravity circuits, and recovery improved. He then introduced new techniques at both mines. The recovery of sulfide minerals, which react like metals, was enhanced by adding the flotation process, which was then just coming into common use.[24] Stannard found that he could even float the copper carbonates, azurite and malachite, which are nonmetallic in character and generally are not amenable to flotation. He treated the copper carbonates with sodium sulfide and calcium polysulfide. As a result, a thin metallic sulfide film formed on the copper carbonates, causing the minerals to react like chalcocite. In effect, he fooled the flotation circuit.[25] He also had another idea for treating the copper carbonate ore. Copper carbonate minerals are soluble in weakly alkaline ammonia solutions. They are also soluble in acidic solutions. The Kennecott ore's dolomite host would react with and consume acid; thus, acid leaching would have been prohibitively expensive. In contrast, alkaline ammonia would not be consumed by reaction with limestone or dolomite, and it could be continually recycled through the copper recovery process. The method was simple in concept but difficult to engineer.[26] Pure ammonia (NH_3) was delivered to the mine on railcars from Tacoma. At the plant at Kennicott, water was added to form aqueous ammonia, which was then transferred, along with a batch of low-grade copper carbonate ore, into leach tanks. After leaching, the now copper-laden ammonia solution was transferred to evaporators, where copper oxide precipitated and gaseous ammonia was redissolved in water to begin the next leach cycle. The copper oxide was sacked and shipped to the smelter.

The amount of copper saved by the ammonia process was significant. In 1920 Bateman and McLaughlin estimated that 25 percent of the Kennecott ore was the carbonate type, and later miners found up to 40 percent. Before Stannard came to Kennicott, much of the carbonate ore had been lost. The combination of improved gravity methods and new recovery processes—flotation, sulfidization of some carbonate ores, and ammonia leaching of others—brought overall copper recovery at the

Kennicott mill to more than 95 percent.[27] Stannard had definitely proven his worth.

The ammonia plant at Kennicott was rare in its day. (Two similar plants had been built in Michigan and Africa.) Employing the ammonia process to recover copper from chemically basic rocks—when acid leaching would be cost prohibitive—remains a viable option, although it is rarely used today. At the Kennecott mines in the Wrangell Mountains, it worked exactly as Stannard had planned.

During his early years at Kennicott, Stannard became very close to Stephen Birch. Already the stage was being set for the time, twenty-five years later, when Stannard would succeed Birch as the head of Kennecott Copper Corporation. Both Stannard and Birch were fond of a paper trail, and Stannard's detailed drawings and plans appealed to Birch. W. H. Seagrave, then general manager of the Kennecott mines, knew the location of every item in the mines and every man in his organization, apparently by intuition. Seagrave, like construction manager R. F. McClellan before him, detested paperwork and went to great lengths to avoid it.[28]

In early 1916 Stannard's position was secure enough for him to make a managerial move, one that rippled down through the staff. Before 1916 W. H. Seagrave was general manager of the two mines at Kennicott (the Bonanza and the newly opened Jumbo) and of the Beatson mine on Latouche Island. DeWitt Smith was Seagrave's mine superintendent at Kennicott; Earl Dunkle was his superintendent at Beatson. Stannard, who began as a mill man, had attained a position essentially equivalent to that of Seagrave. Stannard was general manager of the concentrators (mills) at Kennicott and Beatson. A man named Stadtmiller was mill superintendent at Beatson under Stannard. What happened next is unclear, but by midsummer of 1916, Stannard was general manager of both mines and mills, and Seagrave was out. Stadtmiller replaced Dunkle as mine superintendent at Beatson. DeWitt Smith left Kennicott at about this time; by 1917 he was at the United Verde mine in Arizona. David Irwin, who was Dunkle's and Smith's classmate, stayed a few months longer. By late 1917, Irwin was mining for Phelps-Dodge Company, a long way from Kennicott.[29]

Dunkle, as a strong Seagrave partisan, decided to leave Kennecott, although he could have returned to his field engineer position. If it had been a matter of popularity, Dunkle would have beaten Stannard. Stannard never had the common touch. He was often at odds with the miners and other engineers, including L. A. Levensaler and longtime Kennecott treasurer Carl Ulrich. It was at this time that Dunkle's miners at Beatson offered to go on a protest strike so he could retain his position there. Dunkle, however, did not want to divide the company. He left, along with Seagrave and mine foreman Melvin Heckey, to mine copper at Contact, Nevada, a few miles south of the Nevada-Idaho border.[30]

Dunkle's decision to leave Kennecott in 1916, triggered by the dismissal of Seagrave, was probably a critical one to his family and to his career. In the years between 1910 and 1916, Kennecott had about as much invested in Dunkle as a future mine manager than as a scout or explorer. Dunkle's first Alaskan job, at the Beatson, was mine based, and his assignment at the Midas project was essentially mine development. It appears that W. H. Seagrave was trying to move his protégé into mine management rather than exploration, options that had personal implications. A mine manager's life is appreciably different from that of an explorer. In those days, each position probably had at least a twelve-hour-per-day schedule. As the mine superintendent worked up the company ladder, his wife moved up the company social chain into a more enjoyable life. The explorer's wife was often left behind. After Florence Dunkle gave up her early independence, she was primarily a housewife and mother and was often lonesome because of her husband's exploration schedule. A young Bill Dunkle, the Dunkle's second son, perceived his mother's loneliness: "[W]hen my father was home things were exciting and wonderful but when he was gone my mother's life would shrink to two small kids and the house."[31] Perhaps some of this loneliness would have been averted if Stannard had not dismissed Seagrave and the small Dunkle family had remained at Beatson for several years before moving to larger mines.

Stannard's stated objective in the management change—to increase copper production during World War I—was partly achieved, and his choice for mine superintendent at Kennicott—W. C. Douglass—proved to be excellent. Kennecott's Alaskan mines had their greatest production in 1916, but some of this edge was lost in 1917, when the miners walked out for several months. The strike was exacerbated by Stannard's patrician attitude, which was easily perceived by the miners and others in the copper camp. (One girl on the "line" at McCarthy nicknamed Stannard the "Duke," a name that stuck.) At Beatson, Stannard's choice of managers was not satisfactory. His man, Stadtmiller, could not handle the work and had to be relieved.[32]

Contact, Nevada, the post-Kennicott destination of Seagrave, Heckey, and Dunkle, also had its best year in 1916, when it produced more than 650,000 pounds of copper. Government price controls emplaced in 1917 reduced the copper price by more than thirteen cents per pound, and production fell. It increased somewhat in 1918 before plummeting again at the end of World War I. The Nevada-Bellevue mine at Contact closed, and the miners moved on: Heckey to work with DeWitt Smith at the United Verde mine in Arizona, Seagrave to open a consulting office in California, and Dunkle to head back to Kennecott.[33]

The Dunkles, now a family of four after the addition of their second son, William, could see several reasons to return to the security of Kennecott Copper Corporation. One of them was an anticipated salary of one thousand dollars per month, an amount that was sufficient to buy two Model T Fords every month, with some left over.[34]

6

From Pier 2, Port of Seattle, to Africa

EARL DUNKLE BEGAN THE 1920s with a hike across the Seward Peninsula in Alaska and ended them with a rail trek across the southern part of Africa. Except for one excursion into Nevada with W. H. Seagrave, Dunkle's decade belonged to Kennecott Copper Corporation and Stephen Birch, but more immediately to his earlier nemesis, E. T. Stannard. Stannard, formerly of Kennicott, Alaska, was based in Seattle, Washington, where he had charge of the mines in Alaska, exploration in the Pacific Northwest, the CR & NW Railroad, and Alaska Steamship Company. Stannard's empire and Dunkle's new job were based at Pier 2, Port of Seattle.

The operating linkage between the Alaskan mines, the railroad, and Alaska Steamship Company was essentially the same as that which Dunkle had known in 1916. Dunkle, again field exploration engineer, relied especially on Alaska's steamships to place him as close as possible to where he needed to go on his exploration ventures. The importance of Alaska Steamship Company to the territory of Alaska at that time can hardly be overemphasized. The company, colloquially called "Alaska Steam," was the equivalent of modern Alaska Airlines, a few international air carriers, and at least two marine cargo outfits. After 1932 Dunkle sometimes traveled to or from the conterminous United States by air, but from 1910 until he left Kennecott in early 1930, Alaska Steam was the prime mover for Dunkle and, indeed, for all traveling Alaskans. As a constant and favored traveler, Dunkle knew every captain, every agent, and probably most of the longtime sailors on the line. Some of the men, including agents Willis Nowell in Juneau and Alex McDonald in Seward (and later Anchorage), were among his closest friends.

The steamship line was of critical importance to Dunkle on his first assignment after he rejoined Kennecott in May 1920. The assignment was to evaluate a prospect near Candle on the Seward Peninsula, some thirty-five hundred sea miles north of Seattle.[1] Dunkle arrived at Nome, the peninsula's main port, on the SS *Victoria* on 13 June 1920. The waters off Nome are very shallow, but the vessel was met by dogsleds instead of the usual shallow-draft lighters because ice was still locked to the shore (fig. 18).

Although mining activity had declined since the rich early days at Nome, there was still much to see and speculate on. Nome was awash with rumors that Wendell P. Hammon, the California dredger king, had acquired the rights to a cold-water process that would thaw permafrost inexpensively, allowing large-scale mining of the frozen low-grade gold deposits left by early miners. If this were true, Nome expected Hammon to bring one or more of his big gold dredges north.[2] Nome was definitely bustling in early June because the *Victoria* was the first ship to arrive and open the shipping season.

Preparation for the property examination took several days to complete. During his short stay in Nome, Dunkle found it difficult to sleep. The walls of his hotel room were cardboard thin. To make matters worse, there was a gap of several inches at floor level, caused by irregular thawing of the permafrost foundation of the building. Moreover, the adjacent rooms were filled with angry Russians, who caroused at all hours of the night and slept during the day. The Russians had more than usual justification to be angry. They had planned to mine on the Siberian coast, but their gear was at St. Michael at the mouth of the Yukon River, waiting for the shore ice to break. They also discovered that the Russian Revolution had finally reached Russia's Far East and that their prospects were overrun with White and Red factions, who were fighting over possession of the Siberian coastline. The circumstances made the location of their mining equipment moot.[3] Dunkle was glad to get out of town.

The next significant part of Dunkle's trip was a long but uneventful hike. Dunkle and an owner of the targeted mining property, I. W. Purkeypile, left Nome by boat on 26 June for Koyuk, a village at the extreme southeast side of the Seward Peninsula (see fig. 6). The men continued by boat up the Koyuk River, which is navigable to Dime Landing, below the mining village of Haycock. From Haycock it is about sixty cross-country miles to the mine, which Dunkle called the Anderson because A. T. Anderson of Seattle was funding its development. With full packs, Dunkle and Purkeypile walked to the mine in sixty hours, arriving on the night of 3 July. Because of bloodthirsty mosquitoes, the hikers had slept in only fits and starts when it was absolutely necessary.[4]

Dunkle turned the Anderson property down because the deeper level of the mine did not show as much ore as the surface and first level. He

Figure 18. Dunkle on board the SS Victoria *in the Bering Sea, 10 June 1920* (top), *and landing at Nome on 13 June 1920* (bottom). *Courtesy, Dunkle family.*

thought that the miners might have lost the vein in the mine's deeper level, and before he left, he had worked out a plan with the mine boss to crosscut in search of better material. Dunkle also doubted that the lead-based ore, even with the silver that had interested Kennecott, could be delivered profitably to a smelter from its remote location.[5]

The trip back to Nome was more eventful than the inbound trip. Dunkle had considered hiking out through the Inmachuk drainage to the little railway that ran from the Kougarok to Nome. It was, however, a solo one-hundred-mile walk to the head of the railway: in total, at least a four- or five-day trip. Moreover, the schooner *White Mountain,* which was scheduled to reach Nome in three days, was ready to leave Candle on 13 July. The schedule and route should have allowed Dunkle to complete some work at a tin prospect at Lost River on the southwest coast of the peninsula. Dunkle chose the boat, but it would have been easier and quicker to walk to Nome. From 13 to 25 July, the schooner battled full-gale seas in an attempt to round Cape Prince of Wales. Finally the engine failed, and Capt. E. K. Johanson's only choice was to beach the schooner at Lopp Lagoon. The passengers arrived in Nome in early August.[6]

Alaska was much more isolated in 1920 than at present, but there was a good telegraph system. Using company code, Dunkle transmitted confidential information to Pier 2 in Seattle. After a boat trip that greatly exceeded the three-day estimate, a travel-worn Dunkle wired E. T. Stannard, "property scanderait ennuykis of illargiri," which translated as "property is of no value; there is no indication of presence of large ore body." He also transmitted, "I have been delayed by shipwreck. . . . I am to leave on the *Valdez* tomorrow. What are her orders?" Once aboard ship, Dunkle had more time to follow up with correspondence. He again wired Stannard: "We hit the worst spell of weather that even the Eskimos have ever seen in this country and were 22 days getting into Nome, during which we were shipwrecked in Lopps *[sic]* Lagoon on the Arctic and brought into Nome on a whaler. There was no particular danger in the shipwreck, only considerable discomfort and exposure. . . . Unless we are delayed I expect to reach Latouche about the 17th."[7]

Dunkle's telegraphic communication gave Stannard time to consider new destinations and to pass on information or queries about Kennecott's mining competitors. Stannard also told Dunkle that he had tried to get word to Florence Dunkle on the safe outcome of the shipwreck. Some of the company business concerned the Midas property south of Valdez, where Dunkle had driven development workings in 1914. The property had again been submitted to Kennecott for consideration. If Birch gave his go-ahead, Stannard wrote, Dunkle should reexamine the property. If not, Stannard advised, "[Y]ou should take a run around Prince William Sound and out the Government Railroad while weather conditions will permit your looking around."[8]

For Dunkle's surveillance role, Stannard noted that an important gold discovery had been made on Chichagof Island in southeastern Alaska. He would know by the time Dunkle reached Latouche Island whether Dunkle should cancel work on Prince William Sound and proceed to Chichagof via Juneau. The letter closed with, "What have you heard about the Treadwell operations in the McGrath district?" Dunkle received Stannard's letter on board the SS *Valdez* and replied that he was on his way to Cordova, that he had already completed a report on the Anderson property, and that he was aware of the developments by the Treadwell companies. He closed apologetically with "[S]ure lost a lot of time on that trip and am praying for a late fall so that I can clean up everything around here."[9]

Two of Dunkle's projects in the early 1920s were unrelated to the metals needed by Kennecott Copper Corporation. One was a project on paper pulp, a product of potential value for Alaska Steamship Company. The company always sought profitable loads for the return trips to Seattle. As a publicly regulated common carrier, the economic operation of the company depended on a complex mix of passengers and freight items, only one of which was the copper ore shipped to the smelter. For decades the company searched for other bulk items. An obvious target was timber from the southern coast of Alaska, but Alaskan spruce and hemlock could not compete with fir then being cut from the forests of Washington and Oregon. Paper pulp produced from Alaskan trees seemed a better option. Dunkle's role was not to investigate the timber resource, which was assured, but to find an inexpensive and reliable source of hydroelectric power necessary for a pulp mill. His investigations took him throughout southeastern Alaska. Dunkle liked a site on the Spiel River east of Juneau and another site at Thomas Bay near Wrangell, Alaska. He estimated that fifty thousand horsepower was available at either site, which would be adequate to power a substantial operation.[10] Evidently Alaska Steam reconsidered its options because pulp was never developed. The work on hydroelectricity, however, satisfied Dunkle's curiosity, and he used the information as background material on other projects.

Dunkle's second "nonmetallic" project was the final evaluation of the Bering River coal field for Stephen Birch and partners Falcon Joslin and George Hazelet. The evaluation probably took parts of both 1922 and 1923 to complete. Although development of the field had been stymied by the withdrawal of coal permits and leases and by subsequent litigation (see chapter 2, "Alaska in the Good Years"), Birch, Joslin, and Hazelet controlled the field and were eager to develop it, if warranted. Quality and structural continuity of the coal were the issues that Dunkle addressed, and the U.S. Navy wanted to know if the coal was marine steaming grade. The best way to determine that was by test burning a large sample that Dunkle and the crew had mined. The coal failed a burn-quality test on the vessel *Jason,* and the mining venture confirmed the structural complexity of the field. The coal seams were so badly folded and faulted that economic recovery of coal was difficult, if indeed feasible. Dunkle's examination ended a half-million-dollar venture into coal by Birch and his partners.[11]

Dunkle preferred other prospects. He recommended that Kennecott, or perhaps Guggenheim-controlled American Smelting and Refining Company (ASARCO), acquire the newly discovered Lucky Shot gold-quartz vein in the Willow Creek district.[12] Although the prospect received serious evaluation at headquarters, it was rejected. Perhaps the deal that was offered was too steep because Kennecott optioned the Mabel property, a similar gold-quartz vein only a few miles away from the Lucky Shot. At the same time, Kennecott announced the acquisition of the Copper Mountain (Mt. Eielson) deposits in Mt. McKinley (Denali) National Park.[13] The option of the Copper Mountain property, about twenty miles southeast of the historic Kantishna district, allowed Dunkle to have his first look at the country on the north side of the Alaska Range. The existence of two Kennecott options in Alaska justified a Dunkle family move to Anchorage.

The Dunkles moved to Anchorage in the spring of 1923, and Dunkle immediately began work at the Mabel, only sixty miles from town. In late spring he mushed to the more distant Copper Mountain with two young men, Andy Anderson and Slim Johnson, whose assignments were to construct a camp and to haul mine timber from the nearest site, about nine miles east of Copper Mountain. In June Dunkle returned to Copper Mountain, this time with horses. He was accompanied by his foreman, Ira McCoid, several miners, and a cook. In late June Dunkle and Red Grant, one of the property owners at Copper Mountain, made an extended reconnaissance on the north side of the Alaska Range. The two men "siwashed it" (traveling fast and throwing their bed rolls on the ground) as far west as Slippery Creek, into a region being prospected by Bill Shannon, before they headed back east to Copper Mountain.[14]

Dunkle had an appointment in Anchorage in early July. The southbound government-owned Alaska Railroad train left McKinley Station each Tuesday at 1 P.M. It was early afternoon on Monday, and Dunkle thought that he would be hard pressed to make the sixty-mile hike in time to catch the train, so he decided to try a shorter route. The route, which was new to Dunkle, was up Muldrow Glacier, across Anderson Pass, and down the West Fork of the Chulitna River to Colorado Station on the railroad. The route was probably more rugged, but Dunkle guessed that it would save fifteen miles and give him a few more hours' flexibility in meeting the southbound railroad schedule.

The first problem that Dunkle faced was in deciding which pass was Anderson. As Dunkle later wrote to Mt. McKinley explorer and scholar Bradford Washburn:

> There is a false pass a few miles before you come to Anderson. I was following along the east side of the glacier [Muldrow], sometimes on the ground and sometimes on the ice, and was so close to the east mountain wall that I could not tell for sure if that wall continued solidly around the bend of the glacier, or opened up toward the east end and, probably, the Pass. . . . The Little Guy kept me going on up the glacier and I found the true pass, but even then was not sure until I reached the saddle and could look down the length of the West Fork glacier.[15]

The risk associated with the false pass was real. A few years later, McKinley park superintendent Grant Pearson and his chief ranger, Fritz Nyberg, took the false pass by mistake. They were saved because their lead dog refused to approach the clouded two-thousand-foot drop. Dunkle's own adventures were not over. At one point he packed down a runway and broad jumped over a snow-covered crevasse at the top of the West Fork Glacier. Coming off the ice, he found the West Fork of the Chulitna River in flood stage, but he could not or would not turn back. Dunkle had to cross the torrent six times, but he thought that he was in real danger only on the first crossing. He reached the safety of Colorado Station by 3:00 P.M. on Tuesday and had to wait several hours because the train was late. At Colorado Station, he met old friends Lon Wells, Frank Wells, and John Coffey, the discoverers of the Golden Zone deposit.[16]

Dunkle's mining crew continued to explore the Copper Mountain (Mt. Eielson) claims and found deposits of lead and zinc in a complex zone of intrusive *sills* and *dikes* (see glossary) more than two miles long and two thousand feet thick. For many years, both private and government geologists looked with interest at the Mt. Eielson deposits, but the base-metal-rich prospect was too far from the Alaska Railroad to be developed.

The option was dropped. Dunkle mapped and drove development workings at his second prospect, the Mabel mine. The veins were rich in gold, but they were complexly faulted, so Kennecott also dropped its option there. Local miners Charles Bartholf and W. S. Horning took the partly developed mine and recovered a small fortune—nearly sixteen thousand ounces of gold—before 1930.[17]

Although the mining ventures were unsatisfactory, Dunkle's transfer to Anchorage in the spring of 1923 had some benefits for the Dunkle family. Even at Copper Mountain, Earl was only two hundred miles, or two days, from Anchorage. Anchorage was much more civilized than it had been the first time Florence Dunkle had seen it in the winter of 1915 from a cabin at the mouth of Ship Creek. It also was a revelation for the Dunkle sons, Bill and Jack, respectively in the first and third grades in Anchorage's only school. Bill Dunkle remembers that the school did not have enough seats for the first graders, so they sat on the floor on newspapers. Such conditions were not conducive to staying in school all day. Bill's favorite destinations after escaping from school were Sydney Laurence's studio, where he was entranced by the colors that the artist could create from a few tubes of paint and the Brown and Hawkins store, where some small gift, such as a pocketknife, usually awaited him. Third grader Jack Dunkle, always more scholarly than Bill, had a desk and usually completed his days at school before he went home to practice his piano lessons.[18]

The move to Alaska also allowed Dunkle to revisit the Mother Lode at Kennicott in 1924, almost ten years after his last visit. The visit confirmed that his early faith in the Mother Lode project had been justified.

At the end of the 1924 season, Dunkle reported generally negative results to Stannard, although he thought that two prospects—the Liberty Bell near Nenana along the Alaska Railroad and the Big Missouri near Hyder in southeastern Alaska—had merit and could develop into mines (see fig. 6). He evidently planned to return to reconnaissance in 1925 because he mentioned two interesting prospects that he planned to visit.[19] Another opportunity emerged, however: to return to Contact, Nevada, with W. H. Seagrave. Both Stephen Birch and E. Tappen Stannard looked aside tolerantly as Dunkle rejoined his old boss.

Low post–World War I copper prices had doomed mining at the Nevada-Bellevue mine at Contact in 1919. Copper prices had not rebounded significantly by 1925, but other economic conditions had changed. Contact was now linked by railroad to several smelters, including the plant near Salt Lake City. Savings due to reduced shipping costs appeared to be large enough to operate the mine profitably. New backers of the Nevada-Bellevue, including Seagrave and W. R. Rust of the Chichagoff mine in Alaska, acquired the mine and equipped it for production. Total investment on the new project was about $570,000.[20]

Prior mining indicated that the ore was continuous and rich enough to mine. The Bellevue vein could be traced almost two miles as part of a complex swarm of veins. The main vein averaged only about five feet thick, with an average copper content of 5 percent, but there were wider zones and rich shoots of copper-silver ore. From 1913 to 1917, ore at Contact averaged almost 10 percent copper. In 1917, when copper prices began to fall, Nevada-Bellevue produced about twelve thousand tons of ore that assayed 22 percent copper, probably because of high-grading to offset the decreased copper price.

Some ore appeared to be amenable to a new method of mining: in situ acid leaching. Down to the mine's 250-foot level, the ore was a colorful mixture of blue-green chrysocolla, azurite, and malachite; blackish copper pitch; and bright native copper: minerals that are more or less soluble in acid. In 1926 a new level, the Ilo tunnel, was driven at the base of the acid-soluble ore zone. The ore was leached with a 3 percent sulfuric acid solution that was introduced through the old Bellevue tunnel and collected on the new Ilo level. Water for the dilution of the concentrated sulfuric acid, which was brought to the mine in rail tank cars, was pumped eight thousand feet and raised six hundred feet from Salmon Falls Creek. W. H. Seagrave's miners used conventional underground methods in the deeper part of the mine and produced about 1.6 million pounds of copper between 1925 and 1930, when copper prices plummeted to levels never seen before in America. The mine then closed.[21]

Although the mine was not a commercial success, Bill and Jack Dunkle remembered the years at Contact as the most successful for the Dunkle family. Earl was home every night. Ten- and twelve-year-old boys could explore the mines and desert all day and watch their father drive golf balls in the evening. Earl's golf course, across the acid-laden process ditch, lacked scenery but may have supplied some incentive to avoid the toxic trap. As in Anchorage, the Dunkle home contained a piano, which had become a necessity for both Jack and his mother. In the evenings, Earl and Florence played flute-piano duets.

The Dunkle family, especially Earl and Bill, were fascinated by airplanes and aviation. Earl experimented with a model helicopter, which was configured like the military "Flying Banana" with two main rotors. In late May 1927, the Dunkles, along with most Americans, rejoiced when Charles Lindbergh arrived safely in Paris in the *Spirit of St. Louis.*[22]

Although Earl celebrated Lindbergh's flight, he anticipated the failure of the mining project. A few days before Lindbergh landed, Dunkle had appealed to Kennecott through Stannard:

> This leaching we are trying here is working to some extent but not enough to make much on 13 [cent] copper. . . . We could do very well on 15 cent copper or a much lower price on acid but neither seems in immediate prospect so it looks like a shut down until conditions improve. . . . I would appreciate an early reply, for I have pulled myself down to a minus quantity in the effort to pull this property through.[23]

If Stannard chose to reply by telegram, Dunkle advised him to use company code, "for this is a wide open village in more ways than one."

Stannard sent a coded reply. He needed Birch's approval but thought that he could find work for Dunkle at least through the summer season. In his enquiry letter to Birch, Stannard was sympathetic to Dunkle's plight, with some reservations: "I am terribly sorry that Dunkle's mine at Contact has again proven a failure, but as you once stated, I believe it is better that he finally has it out of his system." Among his potential assignments for Dunkle, Stannard needed Dunkle's coal expertise to examine a deposit of oil-rich cannel coal near Nenana, Alaska. Birch, however, had his own ideas about Dunkle's first assignment and had no reservations about rehiring him. Birch wanted Dunkle to look at one property in the States and he wired Stannard: "For the present do not let it be known that Dunkle is in our employ. I have in mind his doing some work and [it] might be best for Dunkle to do it in his own name."[24]

By early June the Dunkles were in Seattle, and Earl was ready to work on multiple assignments: the cannel coal (a fine-grained coal rich in fossil spores), a quartz system at Fairbanks, two properties on Admiralty Island, and one property at Nabesna on the north flank of the Wrangell Mountains.[25] Dunkle's return to the Kennecott fold in Seattle meant his return to an exploration life but also to a stable salary and other compensations for his family. There was a nice home at 8912 Fauntleroy Way SW, with a filtered view across Fauntleroy Cove. There were close companions for Florence and her two sons. Florence's two sisters and their sons, Darwin Badger and Robert Redding, also lived in Seattle at that time. The three Hull sisters—Florence, Marguerite Badger, and Emily Redding—enjoyed visiting, shopping, and lunching together in metropolitan Seattle. It was the first time that all three sisters had been together since Idaho in about 1911.[26]

From 1920 to 1928, Dunkle may have made more prospect examinations in Canada and the northwestern United States than are documented in company records. Dunkle's 1920s address book names mining property owners with addresses that range from California, through Washington and British Columbia, to northwestern Alaska. Some brief descriptions of properties are tantalizing. Frank Mulfatti, of Shungnak and Kobuk in Alaska's Brooks Range, was reported to have a prospect 190 miles upriver from Kotzebue that contained chalcocite ore and could be traced for two miles. The ore was in Hunter Creek, a name not currently in use.[27] After Dunkle rejoined the company in 1927, he renewed his earlier interest in properties on Prince William Sound. Even more than Stannard, Dunkle strongly advocated acquiring those properties and defending them against competition.[28] For Kennecott Copper Corporation, however, imminent copper developments in Africa had a much higher priority than anything at Kantishna, Admiralty Island, or Prince William Sound in Alaska. In January 1929, Dunkle was in New York; by late February, he was on his way to Africa.

Two American companies, Kennecott and Anaconda, played tag with the leadership role in world copper production during the 1920s. Kennecott's mines at Braden, Chile, and Bingham, Utah, could outperform all others except Anaconda's Chuquicamata (usually called Chuqui) porphyry copper deposit in Chile. Chuqui gave Anaconda the edge because it was larger and richer than geologically similar deposits in Kennecott's home territory in Utah, Arizona, and Nevada.[29] Chuqui's grade was about 2 percent copper; the deposits in the United States were beginning to approach 1 percent. By the late 1920s, however, deposits even richer than Chuqui were discovered in southern Africa.

The industrial western world had known of copper deposits in Africa since about 1890, but local knowledge and use of the metal was ancient. The Belgian company Union Minière du Haut-Katanga (UMHK) opened the Star of the Congo mine in about 1910. The mine overcame technical and operational problems and became very successful by 1918. By the 1920s it was certain that a major belt of copper deposits extended in a southeasterly zone from the Belgian Congo (now the Democratic Republic of the Congo) into neighboring Northern Rhodesia (now Zambia). One very important discovery was made as early as 1905 in Rhodesia by hunter and part-time prospector William Collier. In a clearing near the Luanshya River, Collier shot an antelope as camp food for his men. Underneath the downed antelope, Collier found malachite-rich copper ore; he called his discovery the Roan Antelope.[30]

Kennecott and the Guggenheims were well aware of possibly competitive copper deposits in Africa. They had already made at least one attempt to enter the copper belt by trying to acquire the Nchanga mine, but they were repulsed by the London-based Anglo-American Company. A better avenue into Africa might be through Chester Beatty, an American with a base in London but a longtime affiliation with the Guggenheims. (Beatty had been John Hays Hammond's second-in-command in the evaluation of the copper deposit at Bingham, Utah, the Guggenheims' premier copper mine.) In 1926 and 1927, Beatty's geologist, Russell Parker, found another great copper body composed of high-grade chalcocite underneath Collier's malachite discovery at Roan Antelope. Parker's discovery averaged about 3.5 percent copper. The chalcocite-rich ore could be concentrated to a grade approaching 80 percent copper, a grade that had more options for treatment or shipment than the lower-grade or more complex ores of the copper belt.[31]

There was also copper activity, hence possible competition, close to the southern end of the African continent. One potential competitor was a small but rich deposit called Okiep, then in bankruptcy but being viewed with renewed interest by a combine led by American Mining (AMAX).[32]

In 1929 Kennecott's Stephen Birch and the Guggenheims decided to take a hard look at African copper because competition was imminent. The Roan Antelope mine was in construction, and Okiep, formerly an operating mine, could easily be reopened. To make a nine-month-long examination of several African properties, Stannard, Birch, and Murry Guggenheim sent an Alaskan who had exploration, geologic, and operating experience in many types of deposits: Earl Dunkle.

Dunkle left Seattle for New York in January 1929 to prepare for the examination. On 8 February he picked up his "Letter of Indication," which listed his letter of credit number (65189) and corresponding banks in England and Africa. Earl was in London by late February and took the opportunity to see a few sights, which he shared with his family. A postcard to his attorney father showed the Law Courts, more than two hundred years old. The card to his brother Dane noted that "the man who escorted the U.S. Fleet across in 1916 'touched me' for two shillings about here," on the bank of the Thames.[33]

Dunkle left England for Africa on 1 March on the Royal Mail Steamer *Windsor Castle* with but one scheduled stop: the Madeira Islands, the picturesque Portuguese colony. Dunkle steamed into Cape Town, South Africa, on 18 March through Table Bay, one of the most scenic landfalls in the world. On the port side are the shimmering beaches at Blouberg, now marked with luxurious villas. Near the center of the view, Cape Town is at first almost invisible at the base of Table Mountain. Gradually a few tall buildings indicate the presence of a settlement. The mountaintop itself might be mantled by a flat cloud, called the "tablecloth," which appears when wind and moisture conditions are right, as they often are.

On 4 April, after visiting his corresponding Standard Bank and traversing broad and colorful Adderley Street, Dunkle departed Cape Town for Johannesburg on the route of the famed Blue Train.[34] The railway heads slightly north of east through the suburbs of Cape Town, then enters the wine country, which has been intensely cultivated for more than three hundred years, before it diverts north around Du Toits Mountain at Paarl. The route then turns briefly toward the southeast before it becomes serious in its mission, a steady climb to the east-northeast. The countryside retains some coastal greenery in the Hex River Canyon, but at Beaufort-West, the route crosses the escarpment east of the Nueweveld Divide and enters the arid, sparsely vegetated Karoo Plateaus (fig. 19).

The trip into the largely untamed continent with its rich deposits of gold, diamonds, and copper must have been an epic one for engineer-geologist Dunkle. About two-thirds of the way to Johannesburg, the Transorange Route meets the Blue Route. The "Big Hole," the cylindrical open cut on the Kimberley diamond-bearing pipe, is almost at the junction of the two rail lines. In Dunkle's mineralogy class notes of 28 January 1907 at Sheffield School, Dunkle had drawn a cross section of the Kimberley diamond pipe (he spelled it *Kimberly*), one of several pipes "mostly near Kimberly in a circle of three or four miles radius."[35] In 1929 he was there in person.

At Ely and Contact, Nevada, and at Beatson and Kennicott, Alaska, Dunkle had worked with W. H. Seagrave. About thirty years before Dunkle's African trip, Seagrave had been at Johannesburg helping to sink the first deep shafts to penetrate the gold-bearing reefs of the Witwatersrand. At about the same time as Dunkle's trip, foremost economic geologist Waldemar Lindgren noted that the Village Deep mine in the Witwatersrand was about eight thousand feet deep, and "[o]wing to a favorable geothermic gradient, it will be possible to go considerably deeper." Many of the gold mines on the Witwatersrand were then about one mile deep.[36]

Although the Witwatersrand mines were, and still are, the leading gold mines in the world, mature gold mines were not Dunkle's main assignments. Probably the Bushveld Complex, a crudely oval-shaped massif composed of layer upon layer of dark igneous rock capped by red granite, was of greater interest. Dunkle was in the area only five years after a farmer found platinum nuggets on his land near Lydenburg in the Transvaal, a discovery that led to systematic prospecting of the Bushveld

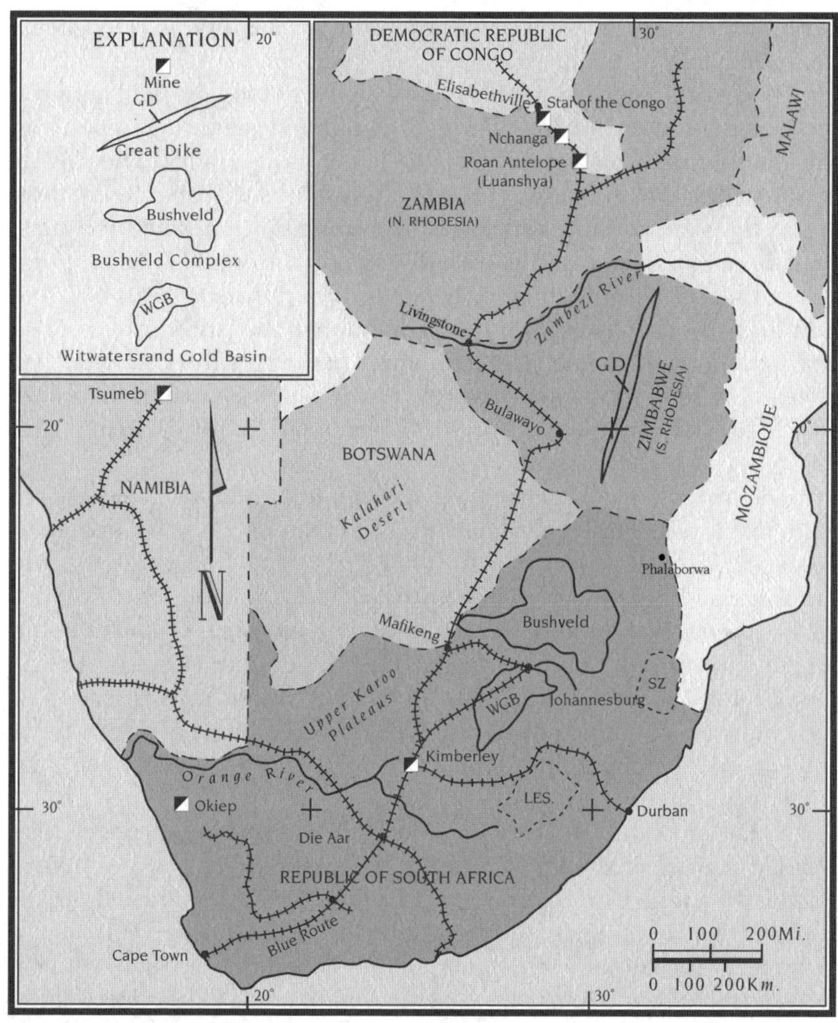

Figure 19. Southern Africa: Rail routes, selected mines, and geologic units. Dunkle traveled from Cape Town to areas as far north as Elisabethville (Lubumbashi), past the diamond pits at Kimberley and the gold mines of Johannesburg. Author's compilation; geology after Dixon, Atlas.

by Hans Merensky and others, which was still in progress when Dunkle arrived in Africa.[37]

Dunkle's main assignments, however, were copper mines, most of which were farther north near Elisabethville (now Lubumbashi) in the

Congo. The copper belt is fifteen hundred air kilometers north of Johannesburg but much farther on the ground. The railway twists around mountains and deserts, connecting mines and remote settlements. From Mafikeng, the site of a major battle in the Boer War, the route skirts the Kalahari Desert until it reaches Bulawayo. It then turns west as far as Livingstone, where the Zambezi River plunges over Victoria Falls, one of the great wonders of the natural world. From Livingstone, the route turns back toward the northeast and gradually swings north to Ndola at the south end of the copper belt. From Ndola, the trend is northwest, linking mines, smelters, and refineries of the copper-rich region.

Dunkle arrived in Elisabethville in late April and by early May was ready to return south to "Rhodesia and the Transvaal," respectively now Zimbabwe and the northeast part of the Republic of South Africa. The trip to the north limit of the copper belt was probably reconnaissance in nature because Dunkle returned to Elisabethville in September. On one, or probably both, of his trips to the copper belt, Dunkle visited his must-see destination, the Roan Antelope (now named Luanshya).[38]

Dunkle received a warm welcome at the mine. Roan Antelope's general manager was his Yale classmate and former Kennecott associate, David D. Irwin, whom Dunkle had not seen since 1916. (After leaving Kennecott, Irwin briefly managed a mine in Mexico before he was named general manager of the Copper Queen Division of Phelps-Dodge Company in Bisbee, Arizona, where he oversaw the mines, smelter, and town, even the Copper Queen Hotel.) Irwin's stint for Phelps-Dodge was an ideal apprenticeship for leading an African venture that required wide-ranging experience and innovative ability. In a later interview conducted by another classmate, H. C. Carlisle, Irwin told Carlisle that he had little direct supervision from London, only a letter every month or so. Irwin's assignment—to build a five-thousand-ton-per-day mine at a location twenty miles by ox team from the end of the railroad—was plagued by disease and rumors. Malaria attacked his expatriate engineers, and rumors of a mile-long snake living in the Luanshya River frightened his native African employees. Irwin persevered. Malaria was conquered through a program outlined by Sir Malcolm Watson, and the snake was finally exorcised.[39] The opening of the Roan Antelope mine in 1931 coincided with the worst copper price in history during the Great Depression. The mine was smaller than the nearby Nchanga and Mufulira deposits, but ultimately it was highly profitable. More than forty years later, in 1971, Roan Antelope was one of the largest copper mines in the world, producing 180 million pounds of copper ore annually from underground workings.[40]

In addition to Irwin, Dunkle had two other Kennecott-era contacts in Africa: A. B. Emery and DeWitt Smith. Emery, who had briefly managed the Kennecott mines in Alaska before W. H. Seagrave, managed a mine in the Transvaal. In Alaska Emery had difficulty managing a mine that was impacted by snow and ice most of the year. In Africa, however, Dunkle was impressed by Emery's well-run operation, especially his creative assaying. Emery's ore consisted of coarse copper-rich bornite that cemented a badly fractured rock, or *breccia* (see glossary). Highly accurate geologic maps and measurements of the visible bornite across wide stopes allowed Emery to control the ore grade without a heavy chemical-assay burden, an advantage in a remote country.[41]

DeWitt Smith arrived in Africa about the same time as Dunkle, but Smith stayed on and probably reached the zenith of his career in Africa. Smith, who had been the mine superintendent at United Verde, Arizona, after leaving Kennecott Copper Corporation in 1916, was probably at the Okiep mine in 1929 as part of the United Verde–AMAX joint venture. Shortly afterward, Smith joined Newmont, the third and also the managing member of the Okiep combine. Because of the worldwide depression in copper prices during the 1930s, Okiep was not profitable for several years. Ultimately, however, Okiep was one of three mines in Africa that helped ensure Newmont's success as a major mining company. (The other two mines were Tsumeb and Phalaborwa.)[42] As a director of Newmont and president and director of South African Copper Company, Smith's expertise was critical to Newmont's ultimate success, especially after World War II, when a fixed gold price made most of Newmont's earlier gold mines unprofitable.

In early September 1929, when Dunkle returned to Elisabethville, he sent several postcards and letters to Florence, his parents, and his brother Dane that crossed paths with other mail. Florence Dunkle made sure that the Christmas presents sent by Bill and Jack to their father left Seattle by Labor Day in order to arrive in Africa by Christmas.[43]

After the May trip to Elisabethville, Dunkle visited Southern Rhodesia and examined several prospects near Bulawayo in the southwestern part of the country. He found large, rich deposits of gold, copper, and chromium. The prospects were sufficiently good to recommend that Kennecott acquire them.[44] At one of the deposits that he visited and endorsed, Dunkle estimated a mineable reserve of more than 96 million tons of chromite ore.

On his return to New York near the end of 1929, Dunkle fought for the acquisition of the properties in Southern Rhodesia and recommended investments in other African mines or mining companies. Kennecott did invest in African mines, but they were unprofitable because of the severe worldwide depression in copper prices, a fate that mirrored that of Newmont and AMAX during the 1930s. Kennecott's acquisitions were ultimately divested.[45] Dunkle could not make progress in his campaign to acquire other African mines. One factor was a disagreement among the top managers on how to maintain Kennecott's prominence in mining, especially copper. Stephen Birch and Murry Guggenheim proposed to sustain Kennecott's leadership by acquiring and operating mines. They were unafraid of competition. E. Tappen Stannard and other directors, who are now unknown, wanted to form a cartel of copper producers to control the price of the metal.[46] Three other factors certainly influenced Kennecott's decision on African mines: Africa was a distant location, far from Kennecott's base in the Americas; Birch and Guggenheim needed cash to acquire more of Nevada Copper at Ely, Nevada; and, while Dunkle was in Africa, the Great Depression began, leading to global uncertainty.

The 1929 trip to Africa was the culmination of Dunkle's career as an explorer and scout. Thereafter his main efforts were in gold mining. The African trip allowed him to renew friendships with Dave Irwin and DeWitt Smith and to meet miners and Africans whose stories enriched his later life. One acquaintanceship led to a serious personal relationship later. Sometime during the African trip, Dunkle met a young society matron, Mrs. F. A. Rimer, named Gladys but always called by her nickname, "Billie." After graduating from St. Andrews University in Scotland with first-class honors in zoology, Billie emigrated to South Africa in 1925 and taught biology at the University of South Africa. When Earl met her, she had a two-year-old daughter, Diana, and was a Ph.D. candidate in biology at the University of Cape Town (fig. 20).[47] The circumstances of their meeting are unknown, but they must have been memorable because after Florence Dunkle died in 1931, Earl brought Mrs. Rimer to Alaska in 1934.

The year 1929 also marked the end of Dunkle's career with Guggenheim-based companies, and thus, his years with Stephen Birch. On his return from Africa, Dunkle planned to go into business in New York with a Toronto-based engineer, Fred M. Connell. Perhaps to keep a promise that he made to Florence before he left for Africa, Dunkle planned to stay home and manage the office while Connell traveled and sought mining properties. But neither Dunkle nor Connell had the financial resources to withstand the Great Depression without external financial support.[48]

Dunkle believed that he had an ace in the hole, a little mine named the Lucky Shot in Alaska's Willow Creek district. The mine needed

Figure 20. Billie Rimer at Cape Town with her daughter, Diana, 1930. Dunkle met Billie somewhere in Africa in 1929 and married her in Alaska in 1935. Courtesy, Dunkle family.

redevelopment, but the ore that had already been mined was very rich. If there was more of it, the ore would quickly recoup the relatively small amount of capital needed to reopen the mine. Dunkle found the capital at Pardners Mines in New York through entrepreneur Harold E. Talbott and Talbott's mining engineer, Jack Baragwanath.

7

Pardners Mines and the Lucky Shot

HAROLD ELSTNER TALBOTT (FIG. 21), BORN IN 1888 IN DAYTON, OHIO, was intensely interested in the commercial and military applications of aviation throughout his life. An early associate of Orville Wright, Talbott manufactured warplanes for World War I, oversaw the production of aircraft during World War II, and ended his aviation career as President Eisenhower's secretary of the Air Force. In the mid-1920s, Talbott had an idea that may have seemed out of character and discordant with his times. He decided to form a mining company to look for gold and other metals in remote parts of the world. On its face, the idea appeared to be ill conceived. The price of gold in the United States was set at $20.67 per ounce in the 1830s, and it remained at that price for almost a century. Especially after the inflation driven by World War I, gold was not an attractive investment. Talbott reasoned, however, that undeveloped gold properties rich enough to be profitably mined were scattered throughout the hinterlands. Tin, nickel, and other minor metals also had economic potential. Talbott called his company Pardners Mines.

Talbott had personal assets. Although he was only thirty-seven years old in 1925, he also had extensive business experience, some of which was acquired in remote regions from assignments with the family-owned Talbott Company. The company had been formed by Harold's father—also Harold E.—to build railroads, docks, and processing plants for resource industries in southern Canada. Young Harold had begun a construction apprenticeship at the age of eleven, when Talbott Company constructed a railroad and docks north of Lake Superior for Algoma Steel Company. Talbott was an experienced northern construction hand even before his college (Yale) years, a fact that may explain his optimism for his new company's chances in the byways of the world. In 1916 Talbott's interest in aviation moved to the fore when he joined with Orville Wright to form Dayton Wright Airplane Company to construct warplanes. After

Figure 21. Harold E. Talbott, ca. 1936. Talbott was chairman of Pardners Mines of New York. Pardners furnished backing to Dunkle at the Lucky Shot mine and other projects. Talbott and Dunkle also shared a passion for aviation. Talbott's aviation career began in World War I and ended in the mid-1950s, when he was President Eisenhower's secretary of the Air Force. Copyright Dayton Newspapers, Inc., all rights reserved. Reprinted with permission.

World War I, Talbott set up business in New York and was chairman of the board of Standard Packaging Corporation.[1]

To break into mining Talbott needed specialized knowledge, something that, at that time, could be acquired in New York. Talbott's first choice for resident mining expert was John Gordon ("Jack") Baragwanath, whom he hired as president of Pardners Mines in 1928. Somewhat later, Talbott hired another experienced mining engineer, Philip D. Wilson, as the company's vice president. To extend the scope of his small New York staff, Talbott used experienced consulting engineers, one of whom was Henry C. Carlisle, a 1908 Sheffield classmate of Dunkle.[2]

Jack Baragwanath, a 1910 graduate of Columbia University, brought several assets to the new company. He had experience with commodities ranging from gold to vanadium, and like Dunkle, Baragwanath could double as engineer and geologist. He had a good mining reputation, which he had largely made between 1912 and 1919, when he had worked in Central and South America for Cerro de Pasco Corporation. From observations he had made overseas, Baragwanath developed a theory that there was a critical length of time for an engineer to spend in foreign climes. If the time was too long, the engineer would become an expatriate, perhaps wealthy but without domestic contacts. If it was too short, as Baragwanath remarked in a later memoir, "without having achieved some reputation, he may find himself working as a shift-boss in Arizona."[3] Just before 1920, Baragwanath decided he was at that critical point and returned to New York, where he joined Harold Kingsmill in a consulting practice. In 1922 Baragwanath left Kingsmill to sign on as a staff engineer for ASARCO in New York. He worked for ASARCO until Harold Talbott hired him to manage Pardners Mines.

The arrangement had advantages for both Talbott and Baragwanath. Talbott gained an engineer with wide experience and professional contacts. Baragwanath could choose mining projects but leave most of the detailed business management to Talbott. Baragwanath also had New York–based social and financial contacts, which were assets for a newly formed company that sought speculative financing. Some of those contacts were through Columbia classmates and the mining industry, but many of his more important ones were acquired secondhand through his wife, Neysa McMein. McMein was a noted magazine illustrator. She was also famous for the eclectic entertainments held at her studio and for her ability to not only play but also invent games for the rich and famous. By the mid 1920s, McMein was financially independent, perhaps because of her financial advisor, Bernard Baruch, whose advice she shared with presidents and nations.[4] Her husband, however, earned fame on his own.

Jack Baragwanath—tall and handsome and sometimes sporting a Clark Gablesque pencil mustache—was aptly nicknamed "Handsome Jack" (fig. 22). He probably was the most socially engaging of Dunkle's mining contemporaries and, in the 1920s and 1930s, one of the best known. Baragwanath's unquestionable skills as an engineer and geologist were often concealed by his public posture. He dabbled in art and wrote two mining memoirs and a play. The first memoir, *Pay Streak,* was published in 1936 shortly after the publication of John Hays Hammond's two-volume autobiography (1935). A reviewer for *Time* contrasted the two works: "[A] successful mining engineer now little more than half Hammond's age offered a volume of reminiscence as informal as Hammond's was ponderous, less than half as long and twice as funny." Jack's second memoir, *A Good Time Was Had* (1962), covers some of the same ground as *Pay Streak* but is also the story of Baragwanath's love for, and rather unconventional marriage to, McMein. Baragwanath's play, *All That Glitters,* which starred Arlene Francis and Allyn Joslyn, had a brief run on Broadway, probably the first and only Broadway success by an American mining engineer. One reviewer was initially favorable; he thought that the play "looked as if it might last."[5]

Baragwanath's ability as a mining raconteur was probably unequaled in his heyday. It is easy to imagine Baragwanath in the old Mining Club in New York, standing beneath the nude by McMein that graced the bar,

Figure 22. John Gordon ("Jack") Baragwanath, mining raconteur and president of Pardners Mines. A good geologist and engineer, Baragwanath was perhaps best known as the husband of glamorous illustrator Neysa McMein. Courtesy, Jenny Lind, sketch after photographs in Time *(1936) and* Mining Journal *(ca. 1946).*

with a Glenlivet in hand, spinning yarns. Baragwanath had a wild story concerning a toucan that always convulsed William S. Paley of CBS fame and brought threats of excommunication from Neysa.[6]

Baragwanath, however, was serious about his mining affairs and quickly recognized two mining opportunities that fit Talbott's criteria. One was a gold placer deposit at Bulolo, New Guinea, a project that might also benefit from Talbott's expertise in aviation. The other was a tin placer in Malaysia. Both projects were ultimately successful. By 1962 the Bulolo enterprise, constructed from air-transported pieces of a large gold dredge, had produced about $75 million worth of gold, had paid nearly half that much in dividends, and had enabled the early growth of Placer Development Company, the predecessor of present-day Placer Dome, one of the world's largest mining companies.[7] In 1930, however, before any dividends arrived in New York from Bulolo or Malaysia, Pardners Mines needed a profitable project immediately to survive. That project was supplied by Earl Dunkle, and it was in Alaska.

In the summer and fall of 1929, Dunkle was in Africa, and Baragwanath was in Europe. On their return home, after the stock market plummeted in October 1929, everyone was looking for financial shelter. Dunkle was still in New York, briefing Kennecott on the African trip in late 1929 and early 1930. As he had done in the early 1920s, Dunkle suggested that Kennecott acquire the Lucky Shot mine in Alaska's Willow Creek district. There is good evidence that the proposal also went to the Guggenheims (ASARCO), but they, too, turned it down. Dunkle was in

E. T. Stannard's office in New York when Cap Lathrop, the wealthiest Alaskan, came through town. Dunkle suggested that Cap take on the Lucky Shot, but Cap also declined.[8] At this point, according to Dunkle, Kennecott's treasurer, Carl Ulrich, suggested that Dunkle contact Pardners Mines. In his autobiographical memoir, *A Good Time Was Had,* Baragwanath says that ASARCO told him that Dunkle had a prospect worth Pardners's attention. Baragwanath described meeting Dunkle and later recommending that Pardners option the mine:

> In 1931 [*sic;* it was 1930] something happened that saved Pardners Mines. Dick Goodwin of A. S. & R. [ASARCO] telephoned one day that he was sending up W. E. Dunkle, a mining engineer who was trying to raise money for a little gold mine in Alaska, a property that was too small for Goodwin's company but might fit our picture. Dunkle proved to be a big, fine-looking man of about my age, a Yale graduate, who had worked for years in the employ of Kennecott Copper and had had wide experience. He told his story briefly, convincingly and with no sales-talk. A Canadian named Thompson had a little company called Willow Creek Mines whose only asset was the Lucky Shot gold mine. . . . The mill and camp buildings had burned down and Thompson was prepared to give a fifty percent interest in the property to anyone who would put in a new mill and rebuild the camp and the cableway. . . . Dunkle expected a moderate salary plus a reasonable interest in the profits.
>
> Dunkle had just returned from Alaska where he had sampled the Lucky Shot and made some maps. He figured that the mine could be put into operation again for $110,000 and that there was enough ore in sight to retrieve this sum. Any additional ore that could be found, and he thought some could, would be gravy.
>
> I discussed this with Harold Talbott, and he agreed to finance the rehabilitation of the mine.[9]

Baragwanath's account, which was not published until 1962, is one year off but otherwise matches Dunkle's version of the events. According to Dunkle, Talbott's acceptance of the Lucky Shot prospect also depended on a favorable engineering report by West Coast consultant Henry Carlisle. The use of Carlisle as examining engineer did not hurt Dunkle's cause.[10] Carlisle's mining objectivity was not in doubt, but his analysis would be based partly on his assessment of Dunkle, whom he had known since 1905 at the Sheffield School. Carlisle visited the Lucky Shot mine, reviewed the data, and reported favorably on the project. Dunkle was set to begin a whirlwind of activities in the 1930s. It was, however, not the easiest decision for the Dunkles to make.

Although he had been the promoter of the Lucky Shot, Dunkle had some practical reservations about the project, as did Florence. The United States was in a serious depression. Dunkle was well known and well liked in the Guggenheim organizations, and he could have stayed there. The Lucky Shot deal promised a share of the profits for Dunkle, if there were any profits, which is never a sure thing in a mine. The guaranteed salary from Pardners was less than half of his Kennecott salary. Bill Dunkle remembered that his parents agonized over the decision.[11] They decided to mine the Lucky Shot, and the decision paid off for both the Dunkles and for Pardners Mines. Baragwanath, in analogy to one of Neysa's favorite characters on Coney Island, referred to the Lucky Shot mine as the "world's largest midget" and observed, "Well the Lucky Shot, small though it was, turned out to be just that . . . , for during our operation . . . it produced some $5,000,000 in gold."[12] In millennium dollars, Lucky Shot's total production from 1920 to 1942 approximates $75 million. Dunkle produced the largest part of that from 1930 to 1938.

The Willow Creek mining district, home of the Lucky Shot mine, is in the Talkeetna Mountains, fifty air miles northeast of Anchorage. It is the third largest historic gold-lode district in Alaska. From 1898 until 1965, Willow Creek produced more than 650,000 ounces of gold, closely following the Chichagof district's 770,000 ounces but far behind Juneau's 6.8 million ounces.[13] Although Juneau advocates might disagree, the Willow Creek district may rank first in scenic impact. Remnants of old mines hang precariously on razor-sharp granite cliffs. Arêtes hug jewel-like blue-green tarns above broad glacier-sculpted valleys. Independence Mine State Park, with restored and collapsing historic mine buildings, is now a popular tourist destination. Snow lasts until midsummer on the peaks, but about the Fourth of July, state-owned snowplows open the winding road that crosses Hatcher Pass, the divide between the Little Susitna River and Willow Creek, the location of the Lucky Shot mine.

Much of the early history of the Willow Creek district revolves around Orville G. Herning and the Bartholfs: Byron, Charles, Eugene, and William. In the earliest years of the district, Herning consolidated and mined deposits of placer gold in the gravel of Grubstake Gulch, a tributary to Willow Creek. The Bartholfs also held placer ground, but more importantly, they were a one-family Willow Creek lode-prospecting crew. Between them, they discovered all the major lodes of the district, except the Fern and the first discovery, Robert Hatcher's Skyscraper vein.[14]

Skill seems to have accounted for all of the Bartholfs' discoveries except the last and best. In 1907 William and Eugene discovered the Granite Mountain vein, the main vein of the Independence group. The same year, William discovered the Gold Bullion. In 1911 the brothers struck gold in the Mabel vein. A discovery by Charles and Byron added the Gold Cord to the family total in 1915. In 1918 a party of ptarmigan hunters was on the north side of Craigie Creek, east of its junction with Willow Creek. Legend has it that a successful shot aimed at a bird that was perched on an outcrop led the party, which included Charles Bartholf, to a quartz vein glistening with gold; thus the discovery was named the Lucky Shot. A second rich vein outcrop was found east and about five hundred feet below the Lucky Shot. It was called the War Baby. Later underground workings proved that the War Baby vein was an extension of the Lucky Shot thrown downward by the Capps Fault.[15]

Dunkle had watched developments in the Willow Creek district since his first visit there in 1912. The Lucky Shot–War Baby appeared to be the best of a series of good discoveries. Access to the district was much better than it had been on his first visit. The Alaska Railroad had a station at Willow, which was only eighteen miles by wagon road from the Lucky Shot. Dunkle shared his enthusiasm about the Lucky Shot with L. C. Thomson, a druggist from Montreal, Canada, who had already managed a successful mine in the Willow Creek district. In 1916 Thomson had leased the Bartholfs' Gold Bullion claims and formed Willow Creek Mines, Ltd. to operate the mine. The Gold Bullion was already the largest mine in the district, but production increased under Thomson's management. Total production at the Gold Bullion was about seventy-seven thousand ounces of gold, an output that exhausted the mine. The ore deposits at the Gold Bullion had formed in nearly flat-lying veins that had almost no rock cover and were exposed near the top of a gently sloping mountain. It was a good example of "what you see is what you get." Although the mine continued to produce on a small scale until 1927, it was clear by 1919 that the supply of ore was very limited. Willow Creek Mines would have to acquire another mine to continue operations. The Bartholfs' Lucky Shot, only two miles away, proved to be the answer.[16]

Thomson began intensive development of the consolidated Lucky Shot–War Baby mine in 1922. He had two main projects, one on each side of the Capps Fault. Thomson started a main haulage tunnel (*adit,* see glossary) about five hundred feet below the original Lucky Shot discovery west of the Capps Fault. The tunnel, a crosscut completed in 1923, intersected both the vein and the footwall of the Capps Fault. The workings were continued to the west as a drift on the vein. Thomson's miners drove a shorter crosscut into the War Baby segment east of the fault and drifted westward until they hit the hanging wall of the Capps

Fault. During mine development, production continued from the Lucky Shot at the two-hundred-foot level and in the shallow Bartholf workings.[17]

The drives toward the Capps Fault from both the hanging wall and footwall sides guaranteed the early success of the mine. The fault bisects and displaces a large and very rich ore shoot. Except for the Fern vein and a few sections of the Gold Bullion and Independence veins, most veins in the Willow Creek district were only two to three feet thick. In the ore shoot bisected by the Capps Fault, the Lucky Shot and War Baby vein averaged about six feet and occasionally reached fourteen feet in thickness. The ore was rich, about two ounces of gold per ton. Between 1921 and a mill fire in 1928 (there was also a mill fire in 1924), Willow Creek Mines recovered about fifty-five thousand ounces of gold from thirty thousand tons of ore, much of which had been mined near the Capps Fault.[18]

By 1927 most of the ore developed by Thompson's early workings was mined, especially on the War Baby side of the fault. The potential for new ore, however, was evident to several participants in the venture: Thomson; Dunkle; Fred Connell, one of Thomson's investors; and a former Kennecott engineer, J. L. McAllen, who had supervised the Willow Creek Mines operations for years. While Dunkle was in Africa in 1929, McAllen mapped and sampled the mine with the view of establishing a larger one-hundred-ton-per-day operation. The depression, however, effectively stopped McAllen's plans for reopening the mine, and Connell dropped out because his examining engineer did not like the project.[19]

Dunkle arrived at the Lucky Shot in March 1930, probably for the sampling visit that Baragwanath had described. This fairly hurried trip had one surprise for the locals. As reported by the *Fairbanks Daily News-Miner,* Dunkle's "visit to the property has nothing to do with the Guggenheim interests with whom he has long been connected as mining scout."[20] Dunkle returned in May ready to begin mine construction, which started immediately. Logistics for the project were explained by the Fairbanks paper:

> An important part of the big development project will be the construction of a road down Willow Creek from the mouth of Craigie to facilitate the movement of mine timbers from the St. Pierre camp to the mines. The road will be about six miles in length and eventually will be extended to the railroad at Houston . . . , giving the Craigie Creek section a water-grade highway. . . . The road will be a cooperative project, built by the Alaska road commis-

sion, with a substantial part of the cost being borne by the mining company.[21]

Besides L. C. Thomson, who remained an owner, the newspaper did not name Dunkle's new financial partners, but Dunkle was noted as general manager of the operation. Dunkle, with Carlisle's report in hand, probably traveled to New York in April to get final approval from Talbott before returning to Alaska to begin the reconstruction of the Lucky Shot mine. Reconstruction was completed in late 1930 by a crew of about fifty men.[22] Possibly some ore continued to be milled at the old Gold Bullion stamp mill while the new mill was being built at the Lucky Shot; Bill Dunkle remembered that his father, carrying a backpack loaded with gold-rich retort sponge from an amalgam circuit, hiked cross-country across the range to Wasilla in the summer of 1930.[23] Both Bill and Florence Dunkle came up from Seattle to spend the summer with Earl and his miners, occasionally avoiding the thick breakup mud by riding in an Athey wagon (fig. 23).

The scenery near the Lucky Shot mine is so grand that an upward or distant view forgives even breakup and its pervasive mud, which was transformed into wildflower-covered tundra within a few weeks. The view toward the newly constructed mine was impressive (fig. 24). Bill Stoll later described it this way:

> A traveler going west over Hatcher Pass in 1931 . . . could in the distance see a cluster of dark-colored, gabled buildings and the broad gently-sloped roof of a new mill. . . . Below the camp, flat, white, tailing ponds. . . . Above, two sets of aerial tram cables swept down the mountain from mine-portal buildings, one above the other, converging on a terminal at the upper end of the mill. If not hidden by low cloud, waste-rock dumps, light in color against the darker country rock, could be seen cascading down the slopes. . . . Off to the right, much lower on the valley wall, a third mine portal and a waste dump below it would be easy to miss. From this portal, a third set of tram cables could be seen heading at a much lower angle toward the mill.[24]

The camp had three buildings of three stories each (two forty-man bunkhouses and a combined kitchen and mess hall) and small houses for the mine staff, all connected by covered walkways. Miners climbed to the workings on steep but protected stairways. There were storehouses and an assay building. The mill was diesel powered, and waste heat from the diesel engines heated water that was circulated to heat the mine buildings. A hydroelectric plant developed about twenty-five additional

Figure 23. Florence Dunkle with her son Bill, avoiding the mud in an Athey wagon at the Lucky Shot mine, 1930. Courtesy, Dunkle family.

horsepower during the summer season.

Perhaps the most important technical development at the Lucky Shot camp was the mill. It is variously described as thirty-five, forty, or fifty ton, but regardless, it was by far the most modern concentrating plant in the Willow Creek district and perhaps in Alaska at that time. Dunkle's mill men, Leo J. Till and George Rapp, working closely with A. J. Wenig of the Colorado School of Mines, developed the process and erected the mill. The first innovation was the substitution of a ball mill for the stamp mill previously used throughout the district. As in other Willow Creek mills, most of the gold was recovered by amalgamation, following gravity concentration of the ore on gently sloping shaking tables. The gold-bearing sulfide minerals in the ore were concentrated by flotation and reground in a small ball mill for tank leaching by cyanide. Early cyanide plants in the Willow Creek district percolated a weak cyanide solution down through coarse tailings produced by the stamp mills, which were essentially mechanized mortars and pestles. At the Lucky Shot, however, the tailings discharged from the ball mill were too fine grained to permit gravity percolation of cyanide. The fine-grained tailings were agitated to allow percolation of the cyanide and solution of the gold, which then, as in earlier district practice, was recovered in zinc boxes. Mill recovery at the Lucky Shot approached 95 percent, about 15 to 20 percent more than the amount of gold recovered by the early mills in the district.[25] Previously the Willow Creek mines had shipped their gold-bearing sulfide

Figure 24. Lucky Shot mine, the most important and richest mine in the Willow Creek mining district; completed mine and mill complex, ca. 1932. Courtesy, APRD, Lulu Fairbanks collection, P68-69-1542.

concentrates to the smelter at Tacoma, and thus faced shipping and smelting charges.

As the Lucky Shot developed systematically with an eighty- to one-hundred-man crew, the ore delivered to the mill was worth the extra effort devoted to gold recovery. In 1932 ore arriving at the mill averaged 2.79 ounces of gold per ton; in 1933 it averaged 2.65 ounces, and in 1934 it averaged 2.45 ounces. More than a ton of gold—36,194 ounces—was recovered in 1933.[26] Each day, the little mill produced about one hundred ounces of gold from about forty tons of ore. (The huge Alaska-Juneau mine in southeastern Alaska produced ten times as much gold every day but from more than twelve thousand tons of ore, or three hundred times as much ore as the Lucky Shot required.)

With grades and production like this, the Lucky Shot mine attained more than just local fame. Following the Alaska-Juneau mine, the Lucky Shot was either the second or third largest lode gold mine in Alaska, depending on the yearly output of the Hirst-Chichagof mine near Sitka. It was well ranked nationally—Dunkle thought about eighth—and thus, was a good target for acquisition. In 1934 Gen. A. D. McRae of Vancouver, British Columbia, arrived in the district with consulting geologist Ira B. Joralemon in a search for gold mines. The Lucky Shot was not for sale. Dunkle, however, introduced McRae and Joralemon to Charles Bartholf and others in the district with inactive mines. McRae almost bought into the Independence mine project, but Dunkle's success at the Lucky Shot had made the price too high, and McRae and Joralemon left the district for less costly targets.[27]

At the Lucky Shot mine, Dunkle built an excellent staff of engineers and a cadre of miners, who followed him for the next decade. Dunkle's mine superintendent was George Ulsh. His mill men, Leo J. Till and George Rapp, could solve almost any problem at the concentrator. Dunkle hired a young Yale graduate, Jim Stanford, to help with the engineer's chores. With this group of men in place, Dunkle could afford to spend some time in Anchorage, Wasilla, or his home in Seattle.

In April 1931 Dunkle was in Seattle with Florence and the boys. Shortly afterward, he had to make a trip to the mine with a mine official, probably either Baragwanath or Thomson. Dunkle would have preferred to stay in Seattle. Florence had been in ill health for several months, and her condition seemed to be worsening.[28] Dunkle's worst fears were justified. Florence Hull Dunkle died on 12 April before Dunkle could return from the mine. She was only forty-six years old. She had outlived her mother's brief years but had not yet reached the rather short life span of her father, Judge Linn Hull. At Florence's death, Jack Dunkle was approaching college age, but Bill was still in his early high school years. Florence's older sister, Marguerite Badger, solved the immediate problem of family care. Marguerite and her son, Darwin, moved from Chicago into the Dunkle home on Fauntleroy Way in Seattle. If the boys could occasionally escape Marguerite's passion for bridge (Marguerite plus three boys made a foursome), it was a good home life during the few years before college lured the boys away.[29]

The year 1931 also saw the depletion of Earl's own close family. Earl's father, John Wesley Dunkle, died in January of that year. Earl's boyhood idol, his older brother, Lester Dane Dunkle, died 16 April 1931, within days of Florence's death. Earl must have visited his widowed mother in Pittsburgh shortly after Dane's death. He was in Chicago on 15 May, and as shown on the postcard he sent to his mother, he had just left Pittsburgh.[30] Always consumed by work, Earl drove himself even harder at the Lucky Shot mine and elsewhere in Alaska during the next few years.

Timing is everything in mining projects, and Dunkle's timing at the Lucky Shot was almost perfect. U.S. gold production languished in the 1920s because the older ore bodies were depleted and production costs rose against the government-fixed gold price of $20.67 per ounce. Costs decreased, however, as the depression hit. Miners and other employees accepted wages that they would have spurned in the unreal boom of the preceding years. Additional events that were critical to the Lucky Shot and other Alaskan gold mines were centered in Washington, D.C. Business investments were stimulated by the formation of Reconstruction Finance Corporation (RFC), a government corporation with lending authority that was authorized in January 1932 during the Hoover administration. The most significant events occurred during and immediately after the "Hundred Days" of the new Franklin Roosevelt administration in early 1933. By executive order, Roosevelt embargoed shipments of gold and silver; future exports or withdrawals of gold from government vaults could be made only under license from the U.S. Department of the Treasury. This order was placed into law in the Emergency Banking Relief Act, which also prohibited gold hoarding and authorized the U.S. Treasury to call in all the processed gold and gold certificates in the country. On 5 June 1933 Congress passed the Gold Repeal Joint Resolution abandoning the gold standard. On 22 October 1933 President Roosevelt announced that he had authorized the RFC to buy newly mined domestic gold and to sell gold and silver on the world market so that the United States could take "in its own hands the control of the gold value of the dollar." On 25 October the government set a price for U.S. mined gold at $31.36 an ounce, as compared with the world-traded price of $29.01 an ounce. By mid-December, the price of gold was $34.06, or nearly the $35.00-per-ounce gold price that was in effect until 1968.[31]

Gold miners, including Dunkle, watched the happenings in Washington, D.C. with an almost consuming interest. Gold miners had previously appealed to the government to allow them to sell newly mined gold on the world market, which offered prices that were almost one-third higher than the price at the U.S. Mint. Some men in Washington were sympathetic. One proposed solution would have allowed miners to sell crude gold ores and concentrates on the world market but to continue to sell refined gold to the mint. That approach was not adopted, at least partly because it would have meant job losses in the gold-processing industry, and the loss of any jobs during the depression was politically unacceptable. For a while it appeared that partly processed gold could be sold overseas. Dunkle arranged to sell gold amalgam (gold alloyed with mercury, an intermediate mill product) from the Lucky Shot to the smelter at Swansea, Wales. In order to protect, safely ship, and perhaps conceal the nature of the product, Dunkle and his mill man, Leo Till, conceived the idea of "canned amalgam" and purchased a salmon canner to put the plan in operation.[32] Before they sold any gold amalgam, however, the government resolved the issue by agreeing to purchase all gold at a slight premium above the world price. The government program gave producers the price they needed to accelerate production. It also continued and strengthened government control over gold, the aim of the administration.

Dunkle was far from the gold negotiations in Washington, D.C., but he was probably better informed than most Alaskan miners. His backers, Harold Talbott and Jack Baragwanath, were in a position to receive information on domestic gold policy through their longtime friendships with Bernard Baruch and Herbert Bayard Swope. Baruch, the ultimate speculator, and his friend and publicist Swope were active in the early New Deal discussions of money matters, including gold policy. They were far more conservative than most of Roosevelt's advisors and favored retention of the gold standard. Although they could not prevail on that issue, they had sufficient influence with conservative Democrats, whose votes the president needed, that their views could not be disregarded. The pragmatic approach that the administration adopted on the gold price and continued mint purchase enabled gold mining to become one of the strongest industries during the depression.[33]

Pardners Mines was satisfied with its receipts from sales to the U.S. Mint and let Dunkle have free rein at the Lucky Shot. Baragwanath visited Dunkle's operation at least three times, one of which was, for Baragwanath, an almost too memorable snowshoe trip. Bush pilot Matt Neminen refused to land on skis at the mine and dropped them off at a distant lake with their snowshoes. Baragwanath later wrote, "Dunk was all right. He was a big powerful man in tip-top condition but I found myself in serious straits. There's nothing like sitting at a desk all day, every day, and getting a work out over a bottle of Scotch every evening to keep one in splendid physical shape." Baragwanath's conclusion was, "You'll finally just lie down in the snow and beg to be shot."[34]

The ore grade at the Lucky Shot fell considerably after the 1934 season, but it was still high enough to operate a very profitable gold mine, which had paid off all of its capital costs within its first few months of operation. In 1935 the average grade of Lucky Shot ore was 0.93 ounces of gold per ton. In 1937 it was 0.87 ounces, and the Lucky Shot still produced one-third of the gold mined in the Willow Creek district. Development of the mine was systematic, pushing ever westward in a

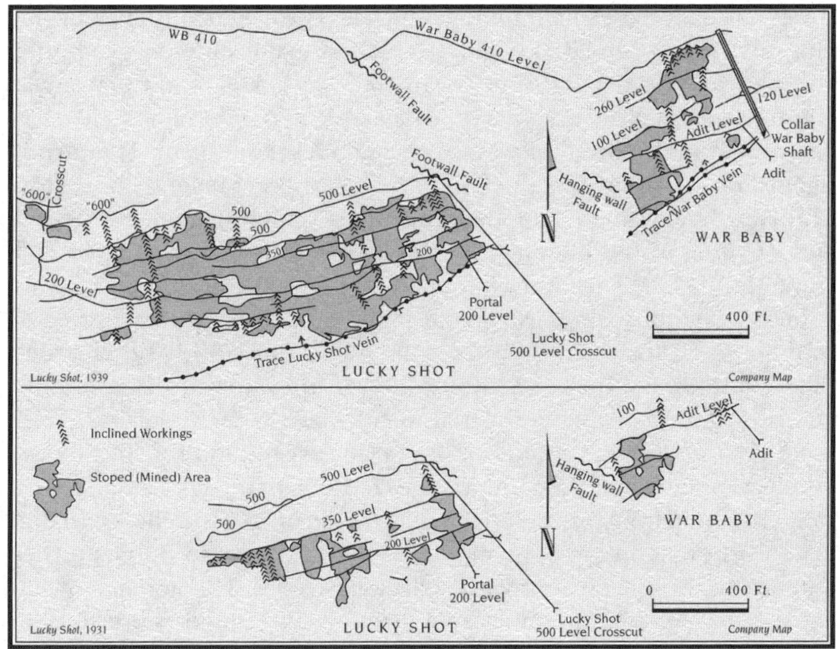

Figure 25. Lucky Shot mine, as developed at the start of Dunkle's tenure in 1930–1931 (bottom) *and at the end of his tenure in 1938* (top). *Author's compilation, after company maps.*

major ore shoot (fig. 25). By 1935, however, it was evident that new ore would have to be found and developed to keep the mine going. The plan developed by Dunkle, George Ulsh, and Jim Stanford called for deep exploration of both the War Baby and Lucky Shot segments of the mine. The project began in 1936 with a five-hundred-foot inclined shaft on the down-faulted part of the vein, the War Baby, and continued with a crosscut to the deep part of the Lucky Shot. The project quickly earned some return from ore in the deep portion of the War Baby, but the results were less promising in the Lucky Shot. Two zones were nearly rich enough to mine, but developing the ore would require cutting intermediate levels above it. George Ulsh favored another option. He proposed to drive through another fault zone at the west end of the Lucky Shot to search for another faulted extension of the vein.[35] Dunkle and his crew, however, lacked the time to pursue either option.

In the spring of 1938, Pardners decided to pull out of the Lucky Shot mine and out of Alaska. There was little doubt that the Lucky Shot could still be mined profitably, even without the spectacular returns of 1932–

1934. But Talbott used his intuition, which had paid off handsomely during the early years of the depression, to make a major move in geography, to Cuba, and in metals, to nickel. In 1938 the winds of war were much more evident in New York than in Alaska. Gold and silver may pay for wars, but iron, copper, chromium, tungsten, and nickel win them: They armor against or cut other metals. Talbott's executive engineers, Phil Wilson and Jack Baragwanath, had identified an extremely large nickel deposit in Cuba that would be much more valuable during a war than a gold mine in Alaska.[36] Few other miners thought as clearly as Talbott. Most miners throughout the great gold-mining states of California, Colorado, and South Dakota kept their focus on gold until World War II struck. The gold miners, however, contributed to the survival of the mining industry at a time when industrial metals were in the doldrums. Their work guaranteed that several of the western states and the Territory of Alaska would escape the worst ravages of the Great Depression.

The company's action took Dunkle by surprise. He believed that he had a firm first-option agreement with its managers to purchase their interest in the Lucky Shot if they decided to sell. The agreement had lapsed, however, and Pardners sold its interest to Dunkle's previous potential partner, Fred Connell, so quickly that there was no time for counteraction. Word of the sale arrived at the mine on Easter Sunday in 1938. The crew, after a more alcoholic party than Dunkle usually permitted, began an exodus from the Lucky Shot the next day.[37]

The Lucky Shot mine had been very good to both Pardners and Dunkle. Dunkle told Luther ("Tex") Noey, a later employee at the Golden Zone mine, that he paid taxes on an annual income of more than $120,000 for three years in a row, presumably 1932 to 1934. Most of the income was from the Lucky Shot mine. In 1937, when the Lucky Shot was in much leaner ore than in its heyday, Dunkle's total income was about $70,000, of which nearly half was his share of the profits from the Lucky Shot.[38] (Dunkle's depression-era earnings have to be multiplied by a factor of at least ten to account for inflation and make a fair comparison with millennium dollars.)

During the same period that Dunkle was managing the Lucky Shot mine, he traveled throughout Alaska looking for and developing other mines. Alaskans of the time wondered how he did it. There are four parts to the answer: Dunkle's share of the profits from the Lucky Shot, which made him financially independent; participation by Pardners Mines in several of the ventures; Dunkle's incredible level of energy; and the airplane. Although Dunkle was not the first Alaskan miner to use airplanes, he was the first to use their full potential to support the mining industry.

8

Alaska's Flying Miner

MINERS EXPLORING A NEARLY ROADLESS LAND were among the first Alaskans to recognize the value of aviation. Perhaps the first mine-related flight in Alaska was on 16 July 1923, when Carl Ben Eielson flew mining attorney R. F. Roth and supplies from Fairbanks to Caribou Creek, a placer mine in the Salcha region. Eielson followed up his first flight when he flew passengers between Fairbanks and the new mining camp of Livengood. In the same year, Moore Gold Company shipped a Jenny aircraft to Nome. The Jenny and the deceptively firm tundra proved to be incompatible, but the pilot, Charles La Jotte, made a few successful flights and left Nome yearning for a better aircraft and more landing fields. Mining was also part of the early aviation scene at Anchorage. Arthur A. Shonbeck, who was the power behind Anchorage Air Transport in 1926, was a miner as well as a city businessman and farmer. Shonbeck's chief pilot, Russel Merrill, was on a supply run to the mine at Nyac in the Kuskokwim region when he was lost over Cook Inlet in 1929.[1]

Dunkle had been looking skyward for several years. His first flight, a one-way trip in 1928 from the Quigleys' mine in the Kantishna region north of Mt. McKinley, confirmed a tentative opinion on the value of aviation. The trip from Fairbanks by railroad and dogsled had taken five days. The return trip from the mine to Fairbanks in a Bennett-Rodebaugh Company airplane took only about two hours.[2] Gradually the romance or pride that had once attended a sixty-mile hike paled, especially when an upward glance showed a possible competitor passing overhead at nearly one hundred miles per hour. Dunkle only needed some time, which he found in the winter of 1931–1932.

Elliot Merrill of Washington Aircraft and Transportation Company at Boeing Field in Seattle was Dunkle's first flight instructor. Earl continued his training with Stephen ("Steve") Mills, one of the two young pilots who assisted Merrill. Mills and the other pilot, Jack Waterworth, also owned the fledgling Northern Air Service, which boasted the ownership of one aircraft. Dunkle had to return to Alaska before he completed his flight training. Mills, sensing a possible opportunity through the wealthy mine manager, asked Dunkle about aviation in Alaska. Dunkle thought that many Alaskans yearned to be fliers and would pay for the opportunity to learn. Furthermore, he offered to grubstake the young fliers if they would move to Anchorage so he could complete his own training. Mills and Waterworth accepted the offer, but the deal became a bit more complicated when Waterworth totaled their aircraft. Mills and Waterworth brought in a former student, Charley Ruttan, to defray the cost of another aircraft, a two-seater Fleet biplane. Mills, Waterworth, and Ruttan arrived with the Fleet on Alaska Steam at Seward on 26 March 1932. Soon after, the young pilots moved into one of Art Shonbeck's apartments in Anchorage and established their flying business. There were, however, too many companies in Alaska with the word *Northern* in their names, so Northern Air Service was renamed Star Air Service.[3] Dunkle kept his word about financing the new company and brought in several of his associates, including Leo Till and Elmer Larson, as additional investors. The Lucky Shot miner continued flight training, soloed, and received license number 27137 as a private pilot on 14 September 1932.[4]

In the early fall of 1932, after only thirty-four hours of flight time, Dunkle purchased and arranged to pick up a new Curtiss-Wright Travel Air in New York.[5] By this time Pardners Mines had made a considerable investment in Dunkle. Although Harold E. Talbott, chairman of Pardners, was the ultimate aviation buff, Jack Baragwanath claims that he tried to convince Dunkle to abandon his new flying career for safety reasons, but Dunkle would not comply. According to Baragwanath, Dunkle first flew to New Haven, Connecticut, for a college football game, where the new aircraft caught fire. Baragwanath vividly described the events that followed:

> This seemed to me a bad omen, and again I tried to dissuade him from keeping the crate. No trouble at all, he [Dunkle] insisted. He'd be in Seattle in four days. When he left he promised to wire me from Seattle. His first telegram, however, was from New Brunswick, N.J. It read: CRASHED HERE LAST NIGHT BUT ENGINE DAMAGED AND WILL HAVE TO WAIT THREE WEEKS FOR REPLACEMENT. AM UNHURT.
>
> The next communication was a wire from some place in New Mexico, which said: CRASHED HERE YESTERDAY BUT ONLY ONE WING BROKEN. WILL WAIT FOR NEW WING AND THEN PROCEED AS PLANNED.

> Dunk crashed again somewhere else before reaching Seattle, so that his trip from New York to Seattle consumed over two months. I wired him: WHY NOT TRY WAGON-TRAIN NEXT TIME? FASTER AND SAFER.[6]

As a storyteller of professional quality, Jack Baragwanath was one of Mark Twain's conscientious historians. Colorado mining historian Robert Brown attributes this statement to Twain: "Many stirring events in history happened to the wrong people in the wrong place, and at the wrong time. The conscientious historian will remedy these defects."[7] Dunkle's plane did not burn at a football game at Yale, nor did it crash in New Mexico. It did, however, crash and stop a football game in Warren, Pennsylvania, which was probably as embarrassing as the fictitious fire at Yale.

The true story is not as brief nor as humorous as Baragwanath's version. It was a lonely voyage with more heart-pounding events. Dunkle's airplane was an open-cockpit, two-seater biplane. Forward visibility past a sturdy Wright J-5 radial engine was poor. Fuel capacity was sufficient for only about three hours of flight. Although the Travel Air was a good performer in the air, it was difficult to handle on the ground. The days were shortening rapidly, and even the daylight sky was darker in the heavily industrialized, coal-burning eastern United States than it is today. Regular airmail routes had beacons, but beaconed air routes were sparse. There was no radio communication for light private aircraft. Airway charts did not exist for Dunkle's route. He used Rand-McNally state highway maps, which showed by a circle if a town had, or once had, an airfield. A white circle on the ground marked the fields themselves. At night some airfields were lit if an inbound aircraft was expected; many others were always dark.

Beginning with Dunkle's first intended refueling stop in New Jersey, finding an airfield was a constant problem (fig. 26). He finally landed at an abandoned but still marked field and hiked over to the highway to find the new location. Dunkle had wired ahead to his uncle, Ed Hendrick, in Warren, Pennsylvania, to expect him in the afternoon, but there were few landmarks over sparsely settled northern Pennsylvania, and he finally had to return to his last fuel stop for better information.

Before Dunkle left New York, he had been cautioned about high-voltage electric lines next to the landing field at Warren. On his second attempt, Dunkle spotted Warren and the airfield, but because he was watching for the high lines and not for the wind sock, his first approach was downwind and too fast for a landing. On his second pass, Dunkle touched down successfully, but deep grass concealed an old road. The brakes locked and the Travel Air flipped over. Dunkle was thrown out of the aircraft and landed on his feet. The propeller was wrapped

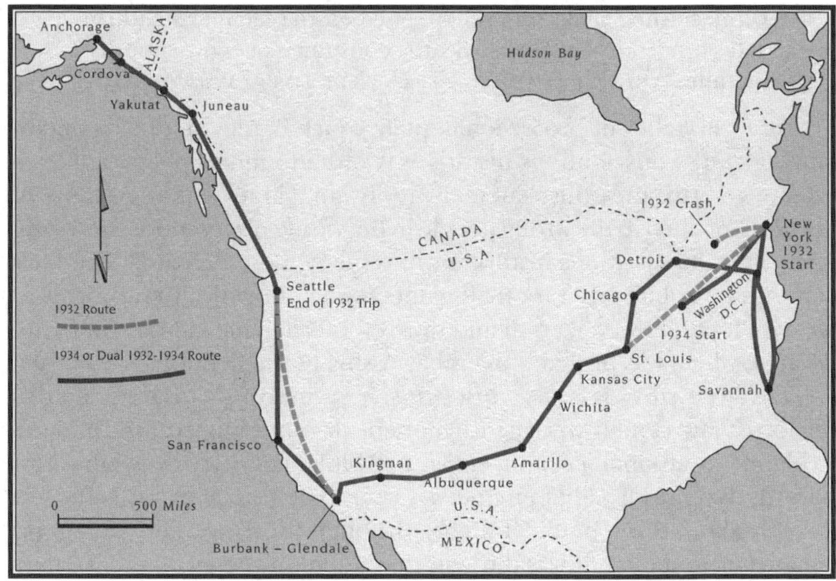

Figure 26. Map of Dunkle's transcontinental flight routes in 1932 and 1934. In 1934 Dunkle put floats on the Waco biplane in Seattle and continued by air to Anchorage in record-making time. Author's compilation.

around the engine, and the fuselage was broken in seventeen places. The crash stopped the high school football game and brought out his uncle and old friends whom Ed Hendrick had rounded up to meet Dunkle. It was an embarrassing reunion after Dunkle's twenty-four-year absence from Warren.[8]

Dunkle spent $2,125 on a new Travel Air ($125 of that amount rented the truck that hauled the wrecked plane back to the factory). On his second transcontinental attempt, Dunkle flew across more populated southern Pennsylvania but still made one emergency landing and an unplanned stop at Johnstown, Pennsylvania, before landing at Harrisburg to visit his classmate Peg Needham. After leaving Harrisburg, Dunkle refueled once in Ohio and intended to refuel again in Indianapolis, where the airfield was on the west side of town. A cold front arrived in Indianapolis at the same time, and when the Travel Air was halfway across the city, the ceiling dropped to two hundred feet. Dunkle found his way out of town by following north-heading streets, dodging smokestacks and towers along the way. He had hoped to find a National Guard field in the northeastern part of the city but had to return to Richmond, Indiana, near the Ohio border for fuel.

Dunkle stopped in St. Louis to install a winter canopy on his open aircraft. Arriving at Kansas City at about 5 P.M., Dunkle sincerely wished that he were down in the heavy automobile traffic as he desperately searched for the field. Fortunately he found it before his fuel was exhausted. Wichita and Amarillo, his next way stations, offered few difficulties. Santa Rosa and Albuquerque, New Mexico, were the next optional landing sites. Santa Rosa was the nearest and more conservative point for refueling, but it was slightly off his flight path. Furthermore Dunkle had been told that the Santa Rosa field would be a tough test for the ground-handling abilities of the Travel Air. He chose Albuquerque, which was near the outer limit of the Travel Air's fuel supply. Crossing the mountain range east of Albuquerque at an altitude of seventy-five hundred feet, Dunkle expected to see city lights. Later he recalled, "All there was, was a river, the Rio Grande River. I had already flown three hours and ten minutes and this ship had only three hours of gas . . . couldn't see the gauge, it was too dark." He thought about using his parachute but reconsidered: "I thought I could save the motor. Also I wouldn't fall on anybody and hurt them. So I decided if the motor konked, I would land in the river." He was saved by altitude. As the Travel Air spiraled downward to the river, the anxious pilot pulled back the throttle to conserve fuel and then spotted a road. He deduced that Sunday evening traffic was inbound toward Albuquerque, but he still could not see the airfield. At the last heart-stopping moment, airfield lights were turned on—someone had heard an aircraft—and Dunkle went straight in without a pass to check the wind direction. On refueling the next day, the mechanic found that there was only about a quart of fuel remaining, probably not enough for a power-consuming go-round of the field to check the wind direction. At about this point in the much later taping of the story, Henry Watkins interjected, "Now that isn't skill, it's foolishness."[9]

Good weather held and visibility was excellent through Winslow and Kingman, Arizona, and into Barstow, California. Conditions were fair as Dunkle crossed the San Bernardino Mountains, but smudge pots warming the orchards below filled the restricted valley with smoke. Forward visibility was almost nil; the only visibility was straight down. Dunkle found Burbank, landed, and then followed the railroad tracks back to Glendale, his destination for a federal inspection of the cockpit canopy installed in St. Louis.

Portland, Oregon, had haze similar to that of San Bernardino. Dunkle held his altitude on the approach to Portland to retain vertical visibility. As he watched, several planes came up through the haze; Dunkle then followed their apparent path in reverse for a safe landing. He landed next at familiar Boeing Field in Seattle, where the Travel Air was disassembled for the marine leg to Seward, Alaska.[10]

Dunkle flew the Travel Air, mostly for short-range scouting trips, for about three years. The plane met its demise when it hit a tree on an attempted takeoff from Lake Spenard in Anchorage. He replaced it with a used Aeronca, a stubby, high-wing, two-seater monoplane. The Aeronca proved to be so valuable for short hops to the Lucky Shot mine and for scouting landing spots for the larger Star aircraft that Dunkle purchased a new Aeronca. He sold the used aircraft to William A. ("Bill") Egan, Alaska's first state governor. The value of small aircraft, such as the Travel Air and Aeronca, for prospect examination emerged shortly after Dunkle's return to Alaska in the fall of 1932. The following summer, Dunkle flew from Fairbanks to Kobuk in the Brooks Range to look at an asbestos prospect and several gold properties. He returned safely in two days from a trip that would have taken weeks by either riverboat or dogsled, depending on the season.[11]

In the spring of 1934, Dunkle made a second transcontinental flight, taking advantage of the knowledge he gained on his first trip. He bought a larger and faster airplane, a four-seater, enclosed-cabin, Waco biplane. The Waco, with a 210 horsepower engine, cruised nearly twenty-five miles per hour faster than the Travel Air and had fuel capacity for six hours of flight.[12]

Dunkle picked up the airplane in Troy, Ohio, and flew it to New York, where he could not resist a flight down Fifth Avenue. He next planned to fly to Savannah, Georgia, to visit his Warren and Yale classmate Will Dunham and Will's wife, Edna, but he made interim stops in Baltimore and Washington, D.C. When Dunkle arrived in Washington, Alaskan congressional delegate Anthony ("Tony") J. Dimond was urging the construction of defense bases, particularly air bases, in Alaska. The proposal had the strong backing of Alaskans, especially because an expanded air-base system would bring long-term benefits for civil aviation in Alaska, one of Dunkle's main concerns. An Alaskan newspaper reported, "It is likely that Mr. Dimond will have the chance to testify at length as to the need for adequate defense measures for Alaska." Although many Alaskans had been to Washington to work with Dimond on the legislation, the newspaper also reported that "not until a few days ago did one arrive flying his own aircraft. It was left to W. E. Dunkle, one of the Territory's leading mining men, to do that."[13] Dunkle left Washington about 1 April 1934 for Savannah. He returned a few days later and encountered some rough weather between Washington and Detroit and near-zero ceilings between Detroit and Chicago.

Dunkle joined his 1932 route at St. Louis. In Wichita he had a chance meeting with Walter Beech, aircraft innovator and manufacturer. Dunkle and Beech had corresponded but had never met. By this time Dunkle and Star Air Service were considering scheduled flights between Anchorage and Seattle. Beech's fast transports were one option for the run. Beech offered to take Dunkle's new Waco in trade for a Beechcraft.[14] Beech and Dunkle took the new Beechcraft out for a spin. The plane's 170-mile-per-hour speed clearly surpassed the Waco's top speed of about 145 and cruise speed of 125. Beech's pilot landed the aircraft at a hot 100 miles per hour, considerably faster than its necessary landing speed and much too fast for Alaska's primitive airfields. Dunkle kept his biplane. Shortly afterward, he was glad he retained the slower, sturdy Waco. At Albuquerque, the wind was gusting to sixty-seven miles per hour across the runway. Following the example of another pilot, who was probably used to the airport, Dunkle landed the Waco into the wind in the sagebrush next to the runway. Mechanics grabbed the wings and helped Dunkle move the plane into a hangar.

From Albuquerque, Dunkle's 1934 route closely paralleled that of his earlier flight, but he did divert to California's Bay Area. Dunkle had arranged to meet L. C. Thomson, an owner of the Lucky Shot mine, and his wife in California. They played some golf, and Dunkle flew the Thomsons over the rising towers of the Golden Gate Bridge at San Francisco before continuing on his journey. At Renton, Washington, Dunkle converted from wheels to floats in preparation for a coastal flight to Alaska. He also picked up Jim Dodson, an experienced aviator from Fairbanks, who was looking for a ride home.[15]

Upon its arrival in Anchorage, the flight was announced as a record for the Seattle-to-Anchorage leg: thirteen hours and twenty minutes actual flying time, within a total time of thirty-one and a half hours. A newspaper reported, "Mr. Dunkle remained at the controls the entire length of the journey." Actually Dunkle had relinquished the aircraft controls for an instant. As they were coming in low over the Copper River flats into Cordova, Jim Dodson spotted a plane heading right for them and grabbed the controls to pull off to the right. Dunkle thought that the planes would have collided without Dodson's quick reflexes.[16]

The Waco was a Star Air Service–Dunkle standby for years. For a time it was Dunkle's favorite aircraft. It flew, as needed, on wheels, skis, or floats. A very young Bill Dunkle, while taking the overflow business from Star, wrecked the aging Waco, first tentatively on the flats north of Anchorage and then more seriously at Moore Creek in the Kuskokwim country.[17]

Dunkle's record flight time was not matched by the first commercial round-trip flight from Anchorage to Seattle, which was made by his own Star Air Service a month later. Steve Mills left Anchorage in early June 1934 with five passengers who had chosen to fly because Alaska Steam was tied up by one of the Great Depression's longshoreman strikes. The float-equipped red Bellanca landed safely at Lake Union in Seattle after seventeen hours of flying. On the return flight, Mills flew solo as far as Juneau, where he picked up L. E. Carpenter, Willie Barber, and placer miner Tony Lindstrom. As recorded by aircraft historians Stephen Mills Jr. and James Phillips, Mills's Bellanca "touched down at Lake Spenard 15 hours and 55 minutes flying time out of Seattle, completing the first commercial round trip flight between Anchorage and Seattle."[18]

Not many weeks after the Star's and Dunkle's record-setting flights, Dunkle left the airways for the railways and seaways of his earlier years. By now a better-than-average pilot, Dunkle was also a widower of about three years. He had a new passport photograph taken at Hewitt's Drug Store in Anchorage. He then left Alaska for London, England, where he met Mrs. F. A. ("Billie") Rimer of Cape Town, South Africa, in what must have been a preplanned visit. Billie had told her husband and daughter that she would follow them to Cape Town on a later steamship. Instead she went to Alaska with Dunkle.[19]

Alaskans became aware of Billie in the late summer of 1934. Willowy, well-dressed Billie, with overlapping Scottish, English, and South African accents, could not be kept a secret in Alaska very long. For Dunkle's sourdough friend Harry Buhro, it was the beginning of a longtime friendship and later a grubstake-mining venture. Russian-born Buhro had spent years before the mast before jumping ship to prospect in Alaska in 1898. Years later Buhro wrote Billie, "The last time we met was in Seattle a few years ago. I well remember the first time, at a bowling alley in Anchorage. Earl had just returned from South Africa and brought you back with him. 'A tall beautiful blonde' a friend told me, 'I just saw her bowling at the alley; she is supposed to be a millionaire of her own.' And so, of course, I went over and met you. A pleasant surprise."[20]

Billie's first visit to Fairbanks was not that public, but it was noticed in the *Fairbanks Daily News-Miner* in mid-September 1934: "Mr. Dunkle tried to fly through Rainy Pass en route from McGrath to Anchorage yesterday and was forced to turn back. He decided to come to Fairbanks with his two passengers, Mrs. G. Rimer and William J. Wagner. The 'flying miner' and his passengers left Fairbanks for Anchorage today in Mr. Dunkle's Waco Airplane."[21] On many of his flights from September 1934 onward, Dunkle had Billie's company in the right seat. It was common knowledge then that Billie was a divorcée, or possibly almost one. All of Dunkle's friends and most other Alaskans were willing to overlook the details. They knew that Earl had been widowed since 1931. One rather strange letter suggests that one person who was not reconciled to the situation was Earl's elderly mother in Pennsylvania. Mrs. Dunkle evidently strongly disapproved of Billie, even after her marriage to Earl and the birth of another grandson. Agnes, who wrote the letter, was a relative or perhaps Mrs. Dunkle's nurse. She thought that no one could have done a better job than Earl on guaranteeing his mother's care, but that did not change Mrs. Dunkle's attitude toward Billie.[22]

At least when he was in the air, Earl was busy enough to forget personal distractions. Flying over Alaska, he did not have to contend with the industrial smog or smudge pots found in conterminous America. At first landing fields were almost nonexistent in Alaska, and the terrain lacked the stubble fields and sage-covered plains that Dunkle had used in emergencies during his transcontinental flights. Alaska had its compensations, however. Skilled pilots, including Dunkle, landed float- or ski-equipped aircraft on mud flats, swamps, and soft tundra, as well as on the water and snowfields for which the floats and skis had been designed. Dunkle thought it was possible to land a floatplane on just about anything. Star Air Service was proud of its ability to land a traveler within three miles of a destination, except in the most rugged mountains.[23]

Dunkle, the pioneer mining engineer and pilot, did have a few advantages that were not shared by all Alaskan pilots. He knew Alaska's terrain from his long hikes, and he had learned to read the weather. Later, when Bill Dunkle flew with his father, Earl was so familiar with the terrain that he did not use maps. Even after Bill was a United Airlines captain, his father always flew left seat at the controls of the airplane. Earl Dunkle had another advantage over many pilots: He was practically resistant to spatial disorientation and vertigo. He was almost never disoriented, seasick, or airsick, even under gale or turbulent conditions. His sons Bill and Bruce, who were commercial aviators for decades, agree with the assessment that their father, like good crop dusters and aerobatic pilots, could withstand unusual attitudes (positions) and forces in the air because of an exceptional balance system.[24] He also never panicked.

Dunkle's pilotage and balance were tested on trips through the Alaska Range, most often in Anderson Pass, the mile-high pass between the West Fork of the Chulitna and Muldrow Glacier in Mt. McKinley National Park. From 1935 to 1942, Dunkle had mining operations on both sides of the range: the Golden Zone on the east and the Slippery Creek, or Caribou, on the west. The pass enabled management of both operations, but it was rarely easy to use, as Earl described to Billie: "If, instead of it costing me about $30 every time it happens, I was given $10 for every time I have had to come back here without getting through to Caribou, I would be a lot of money ahead."[25]

At times, he flew through the pass with frightening results. Jean Porter succinctly described one such trip in *The Flying North,* when she quoted Dunkle: "These dead stick landings are not so good." The trip itself resulted from an earlier crash on a marginal airfield at the Slippery Creek prospect of Bill Shannon. Shannon had promised Dunkle that he would "level off and mark a field for landing, about a mile below the camp on the West Fork of Slippery Creek." Dunkle later reported to Baragwanath that the field was "six hundred feet long and forty feet wide at 3,000 feet altitude, which is not so sweet." By the time Dunkle wrote to Baragwanath, he had already landed there successfully four times. Dunkle and his cat skinner, Al Hamberg, crashed a Stinson aircraft 2 July 1937 on an attempted takeoff from Slippery Creek during bad weather. Dunkle remembered the date because he would not have attempted the flight if he had not promised Hamberg that he could spend the Fourth of July holiday in Anchorage. Dunkle told Jean Porter that the takeoff was uphill.[26] If that is correct, the wind must have been downslope and gusting at flying speed. Landing upslope and taking off downhill are the universal flying rules, unless the downslope wind is strong enough to allow a helicopter-like takeoff. Prudence would have suggested aborting Dunkle's trip.

Dunkle returned to the Slippery Creek field during the following winter with pilot and mechanic Toivo ("Skeezix") Aho and mechanic Herb Engberg to make the downed aircraft flyable. After repair, Dunkle flew the patched-together aircraft. Skeezix and Engberg flew another Stinson and began their flight through Anderson Pass. The trip had elements of excitement for both aircraft. Dunkle explained the adventures many years later to Brad Washburn:

> The motor was very rough because of the bent shaft, and the right wing was so heavy that it took about half an aileron to hold it up, but I figured it would hold together and we started out. The tail end of a heavy southerly storm was still blowing and the air in Anderson Pass was very turbulent. . . . [T]he boys had not drained and cleaned the wing tanks, in fact had only been able to hook up only one of them, and apparently this one contained a lot of ice crystals. . . . The severe agitation stirred up this slush and it blocked off the gas lead and killed the motor. The best looking place for a landing was

> the snowfield of a small glacier to my right, which branched off the main West Fork Glacier near the summit. I got down on it all right, jumped out and double-timed a snowshoe trail crosswise to give Skeezix something to land by.

Skeezix came in and bounced a few times, but he kept the aircraft on the glacier. Dunkle continued:

> On disconnecting the gas line at the carburetor we found the ice indications but had no way to clean the tank, so blew through the line and thus cleared the outlet at the tank. The motor revved up all right, so I took off. The Little Guy must have been in there keeping the ice away from the outlet, because the motor hung on until I got across all of that rough ice of the lower end of the glacier, where there was no place to land, and did not konk out again until I was a couple of miles down on the river bar.[27]

The engine failed again at the Golden Zone mine, where they spent the night in the sawmill camp bunkhouse near the river-bar landing field. It failed a final time over the Goose Bay flats across from Anchorage. Just out of Talkeetna, an aileron wire tore out of a temporary bracket, and the plane began to vibrate badly. Engberg, now flying with Dunkle, reached out through the side of the plane, "which was a little short of fabric at that point," to make a temporary repair.

Dunkle had a more frightening trip through Anderson Pass in May 1937. This time he was alone in the Waco, which was loaded with supplies. Again his audience was Brad Washburn:

> The weather was banked in solid on the Muldrow side and pushing a fairly heavy snow squall through towards the West Fork, which was the side I was on. I stuck my nose into it too far in an attempt to pick up something to fly by over on the Clearwater side of Muldrow, only 2½ minutes away, but I never even got over the near edge of Muldrow, and nearly came to grief due to a very poor piece of judgment. In order to keep something to fly by and thus get through as far as possible I, foolishly, hugged in too close to the black slate mountain on the right side, which is kept clean by the heavy winds which seem to blow there most of the time. Then, when I could go no farther and wanted to get back out of there, I was too close to the wall to turn to the right, which would have kept it in sight, but had to turn to the left into the heart of the clouds and the snow and against the unbroken wall of white of the mountain wall of that side, and I lost contact. It was too close quarters for instruments so I put my face up as close as possible to the windshield and tried to pick

> up something to fly by, at the same time trying to pull the plane around fast enough to pick up the rocks of the right side again, but it was too thick. The turn pulled me too far away and it did not work. Instead, the plane gradually fell off into a sort of a power spin and the motor wound up into a snarl. I throttled back and got things feeling a little better and continued to try to pick up a spot of bare ground.
>
> I knew that time was running out pretty fast and was expecting to crash any second when I happened to pick up one hazy little spot swinging along just under the top rim of the windshield and then another to match it. If the nose had been down one more degree, I would have missed them, because I was coming around rather fast in a steep left bank and the nose was pointed down too much for a proper turn. I kicked right rudder to stop the turn and pulled the nose and left wing up sharply and skidded to the left, just in time to catch the faint outline of the pass, through which I busted out back into the comparatively open weather of the West Fork Glacier, missing the right side of the Pass by fifty feet or less at 140 miles per hour. . . . Anyway, I'll never forget Anderson Pass.[28]

Dunkle knew that split seconds, fractions of a foot, or, as in the case of Anderson Pass, single degrees could be crucial. Whether or not he totally believed in him, Dunkle often referred to a providential "Little Guy" who flew on his shoulder or, at most, no farther than the copilot's seat. Dunkle was not, of course, the first or last pilot to hold these beliefs, but he seems to have had more incidents than most pilots to provide a basis for his faith.

Other incidents in Anderson Pass required neither instant reflexes nor the Little Guy's intervention. They required patience. Updrafts and downdrafts are nearly the rule in crossing the Alaska Range. A downdraft measured on the rate-of-climb indicator at thousands of feet per minute makes anything in the cabin that is not securely tied down into a flying object. One time, when the rate-of-climb indicator showed seventy-five hundred feet per minute, the weightless object was a thirty-pound amalgam retort; another time it was a 350-pound set of dragline chains.[29] Quick action would only exacerbate the problem. The Little Guy probably dictated a firm foothold on the rudder, no increase in speed, and a promise to check the tie-downs better next time.

Occasionally there were better days in the pass. Dunkle crossed Anderson Pass for another Alaska first in April 1937. Although Bill Shannon's Slippery Creek field was barely adequate for a Stinson, a section along McLeod Creek was long enough to take a ski-equipped Ford Trimotor

Figure 27. Waco and Ford Trimotor, Star Air Service aircraft that supported operations at the Golden Zone and Shannon properties on opposite sides of Mt. McKinley, 1937. Courtesy, Dunkle family.

(fig. 27). The *Fairbanks Daily News-Miner* recognized the significance of the flight:

> Demonstrating the future of mining development made possible by aviation, the Star Air Service of Anchorage has flown a McCormick-Deering Tractor from Colorado on the Alaska Railroad, over the Anderson pass, to Slippery Creek in the Kantishna. . . . What aviation means to the mining industry is illustrated by Mr. Dunkle's business, operating from a single base . . . two properties, one on each side of the highest mountain on the North American continent, Mt. McKinley.[30]

In his history of Alaska Airlines, Archie Satterfield stresses the close connection between Star Air Service, Dunkle, and the Lucky Shot mine. The mine furnished most of the dollars that allowed Dunkle's support of aviation. It also gave Star a steady revenue as passengers and freight were flown back and forth between the mine and Anchorage. The Lucky Shot, however, was only part of Star's mining revenue base and a lesser part of Dunkle's vision for commercial aviation in Alaska. Star Air Service also flew for Dunkle projects at the Fairview and Dutch Creek placers in the Yentna district, two prospects in Flat, other destinations in the Kuskokwim region, and the Golden Zone mine. Star also helped to support a Dunkle–

Pardners Mines operation in the Circle district one hundred miles north of Fairbanks.[31] As noted earlier, Steve Mills flew the first commercial round-trip flight between Anchorage and Seattle, and Dunkle or Star flew Tony Dimond on his campaign swings and on other congressional business. These activities required more airplanes and a close circle of like-minded men.

Dunkle; his close business associates Art Shonbeck, Linious ("Mac") McGee, and Alex McDonald; and the founding pilots for Star, Steve Mills, Jack Waterworth, and Charley Ruttan, wanted an airline that could fly throughout Alaska and beyond. As far back as 1924, Shonbeck had been Anchorage's strongest supporter of aviation. Mac McGee seemingly had lesser qualifications. If Shonbeck had not given McGee a job as a truck driver, he might not have survived his first season in Anchorage. McGee, however, turned out to be a good manager of finances and people and had his own firm vision of aviation, one that included radio communication. The fourth member of Dunkle's aviation "kitchen cabinet" was Alex McDonald, who saw the potential of aviation from his longtime position as Anchorage agent and corporate officer of Alaska Steamship Company.[32]

The Star fleet grew, slowly at first. Before Dunkle went to New York in 1932 to pick up the Travel Air, he had bought a Curtiss Robin and put it on floats for Star Air Service. Growth of the company accelerated when Star purchased McGee Airways for fifty thousand dollars. By 1936 Star Air Service owned fifteen of the thirty-nine airplanes that were based in Fairbanks and Anchorage. It advertised itself as the largest airline in Alaska (fig. 28).[33] It was tough to make a profit in a highly competitive market, but air transport had arrived in Alaska.

By 1937, 35 percent of the total freight in the territory moved by airplane, and Star Air Service's big Ford Trimotors and smaller planes moved much of it. The extent of Star's influence is suggested by an early-springtime flight schedule reported in the *Anchorage Daily Times* on 8 April 1935. Because it was spring breakup, Merrill Field was a sea of mud, and all of the Anchorage air fleet had been moved to Lake Spenard, where there was still enough ice for operations: "Four planes of the Star Air Service lined up at Lake Spenard today to take off with loads for long-distance hops. Pilot Johnny Moore is going to Nome, Pilot Roy Dickson to Bristol Bay, Pilot Al Horning to Ganes Creek and Pilot John Littley to the lower Kuskokwim." There were other Star aircraft in the bush: "Pilot Goodman of the Star line is freighting between McGrath and Cripple today. Pilot McLean of the same line is at McGrath." Increasingly Dunkle was drawn into general aviation planning for Alaska.

Travel By Air

Star Air Service, Inc., has the largest fleet of planes in Alaska and is equipped to move anything from a crate of eggs to a tractor.

Planes Meet All Boats at Seward
Fare—$10 each way

•

STAR AIR SERVICE
The Star Way Is the Safe Way

Figure 28. Star Air Service advertisement, Anchorage Daily Times, *1937. Courtesy, Z. J. Loussac Public Library, Anchorage.*

He was recommended as one of four members of the Alaska Aeronautical Board created by the Alaska legislature. The Fairbanks paper reported, "Under the law, the Aeronautical Board must be composed of four persons having a practical knowledge of aviation, one from every one of the four judicial divisions of the Territory." Dunkle received the endorsement of the Third Division, which was headquartered in Anchorage.[34]

Dunkle, Star, McGee, and all of the early pilots and air services also assumed the burden of search-and-rescue operations for downed aviators, a task that today is usually government supported. Occasionally there were humorous overtones to the missions, especially when they were successful, as in the rescue of Anchorage druggist Zac Loussac's supply of medicinal whiskey. One search with no humorous overtones was for Stephen Mills. Mills and a small party of sightseers hit the top of a peak, possibly in a downdraft. The pilot and passengers were killed instantly. There was no warning of an impending crash; Mills's hands were still on the controls.[35] The crash temporarily destroyed Dunkle's belief in aviation. Later reflection on the incident and a comparison with the dangers of pre-aviation life in Alaska gradually led Dunkle to believe that the lifesaving capabilities of aviation offset the tragic deaths of many of Alaska's first pilots.

In 1938, as president of Star Air Lines, the successor of Star Air Service, Dunkle began to reconsider scheduled flights between Anchorage and Seattle. He had nixed the fast Beechcraft airplanes in 1934. However, he could see the merit in the large amphibious flying boats that could fly safely between Seattle and Anchorage. He corresponded with Pan American Airlines about recently surplused flying clippers from its Florida-based run. In the venture Dunkle wanted the backing of both McGee and McDonald, who originally had favored the project. McGee, however, changed his mind. He wanted to go mining. Dunkle did not want to proceed on the new venture without both men's support. He sold his controlling interest in Star to another miner, David Strandberg.[36]

One Dunkle aviation project that is still in full use in Anchorage is the seaplane base at Lake Spenard. Before 1939 Lake Spenard was used as a seaplane base, but it was too small to allow the safe takeoff of heavily laden aircraft. Dunkle conceived of a canal or seaway that would connect Lake Spenard with Lake Hood. A canal would add thousands of feet for takeoffs and landings. Dunkle later called himself the "instigator" of the project, but he could have added "engineer" and "chief lobbyist" to his title. The project enabled the creation of what has been called the largest seaplane base in the world.

Dunkle was not alone in his recognition of the problem. In 1934 two sites, Lake Spenard and Chester Creek, were being considered for Anchorage's seaplane base. City engineer Anton Anderson and some civic boosters backed the Chester Creek site. Moreover, Anderson said that the Spenard site could not be used. Dunkle, however, believed that Chester Creek had more serious problems. In July 1936 the selection process came to a head. A decision was expected from Acting Governor Edward Griffin, the representative of the U.S. Secretary of the Interior, who oversaw the surplus land needed for any solution.

Griffin telegraphed Dunkle about the project on 25 June 1936. Dunkle replied by telegram two days later. On 29 June the governor wrote to Dunkle: "My reference to Lake Spenard project being condemned for consideration . . . was taken from page 2 of said report in which the following appears: 'Lake Spenard has also been used for such purposes but the limit of flooded or flooding area, the location of improved property, roads, etc. prohibits the use or development of this project.' We will now let the matter rest until receipt of your written report and application for the Lake Spenard project, which will then receive our earnest consideration."[37]

In his reply to Acting Governor Griffin, Dunkle made some technical arguments, including the potentially greater length of seaway at Spenard, the ease of construction, and the certainty that the canal would hold water, whereas "the Chester Creek dam may not." He also acknowledged some negatives. "The Chester Creek project is much more convenient because it lies adjacent to the Anchorage Airport [Merrill Field] whereas Lake Spenard is four miles from Anchorage and six miles from the airport." (The negatives would not apply today because the city has engulfed Chester Creek and Lake Spenard.) Dunkle then proceeded to demolish the report that had recommended Chester Creek: "[I]t is a very good report. However, it was prepared by Anton Anderson, who has never had any connection with airplane operation either as operator or pilot. . . . The item that you quote is misleading because it does not refer in any way to the present Spenard project as shown on the enclosed map. I do not believe that Anderson even knew that Lake Hood existed until I sent him out to make the survey on which this map is based."[38]

Dunkle's logic prevailed, but funding was still needed. With introductions from Tony Dimond, Dunkle spent three weeks in Washington, D.C. during the following winter to try to obtain funds for the project. Assistant Secretary of Commerce Johnson was sympathetic but lacked the discretionary funds. Lobbying the territorial legislature proved to be more effective. The Alaska Road Commission reported that, for the fiscal year ending June 1940, an appropriation had been made to construct a one-hundred-by-two-thousand-foot canal, eight feet deep, that would give a total of sixty-one hundred feet for the seaway at Lakes Spenard and Hood.[39] Earl jokingly said that the seaway should be named "Dunkle's Ditch."

Stephen Mills Jr. and James Phillips, in their book *Sourdough Sky*, and the Anchorage Centennial Commission named Dunkle as one of Alaska's one hundred most prestigious early airmen. They said that Dunkle "learned to fly with Steve Mills in Seattle, 1932. Continued instruction and soloed in Anchorage in May, 1932. Owner of Lucky Shot Gold Mine and principal stockholder of Star Air Service. A staunch supporter of aviation, he actively flew his own plane until his death of natural causes."[40] Dunkle had taught Toivo Aho, another of the top one hundred pilots, how to fly. At least twenty-three of the top one hundred flew for Star Air Service at some time in their careers.

This recognition can be extended to the Dunkles collectively, as one of Alaska's first flying families. Earl's position is clear. He was an early pilot and an aircraft company executive. In addition to his correspondence with Walter Beech, Dunkle discussed aviation matters with his friend William Boeing, the founder of the Boeing company of Seattle. As an Alaskan miner, Dunkle was the first to use aircraft to their full potential in the exploration, development, and operation of mines. Although Dunkle had accomplished much of his early work by traveling with Alaska Steamship Company, he quickly recognized the superiority of air transport, except for the heaviest and bulkiest freight. Billie Dunkle, Earl's second wife, was an investor in Star Air Lines. She also understood the importance of the new industry and enjoyed flying right seat on Dunkle's numerous trips. William E. ("Bill") Dunkle, the son of Earl and Florence, began an aviation career as a line boy at Star Air Lines. Bill learned to fly so well that he became a captain, then senior vice president of operations, and finally a director at United Airlines. Bruce Dunkle, the son of Earl and Billie, considered his first love to be geology, but he also flew for decades with United Airlines and retired in late 1996. The oldest son, John Hull ("Jack") Dunkle, followed the law track of the family, but he served in the U.S. Air Force in World War II, where he qualified as a navigator. He also taught navigation and then landed what he considered to be the war's cushiest job: writing navigation instruction manuals and flying on weekly jaunts throughout the United States to stay current.[41]

Wesley Earl Dunkle believed that aviation was part of Alaska's destiny. In commenting on an article by Stewart and Joseph Alsop, Dunkle wrote to the columnists at the *Washington Post:*

> The advent of flying . . . changed the lives and outlook of all Alaskans. No longer was it necessary to put one weary foot ahead of the others for hours on end to go some place. Life itself became less threatening due to the assurance of quick help in case of emergency. All aspects of living were changed in some degree. However, I believe that the greatest change was one of the spirit. It transmitted vicariously to a brave and venturesome population the same feeling of elation which every pilot knows first-hand as he lifts a plane "into the wild blue yonder" and looks about him. Now all was well, and their country would move swiftly towards its great destiny. That is happening.[42]

9

A Sure Cure for Depression

MOBILITY PROVIDED BY THE AIRPLANE and dollars supplied by the Lucky Shot and Pardners Mines gave Dunkle the opportunity to pursue other mining ventures in a period when most of America was paralyzed by economic depression. Primarily he sought gold. For the first time in several decades, U.S. gold was as valuable as that priced on the world market. In Alaska the paucity of docks, railroads, and roads generally precluded development of coal, iron, and copper, which were valuable as bulk commodities. Gold valued at tens of dollars per ounce was one of the few commodities that could be developed almost anywhere it was found in sufficient quantity and grade.

Regardless of his own success, Dunkle was a powerful engine dedicated to keeping Alaska from the economic doldrums of the 1930s. He was assisted by Alaska itself, the huge resource-rich land whose economy tends to be countercyclical to that of the rest of the nation. The depression years from 1929 until 1941 are a case in point. In general, Alaska was much better off economically than the rest of the nation. Although a few Alaskan industries, such as copper mining, were seriously affected (Kennecott closed its Alaskan mines during one period in the early 1930s), others prospered. Alaska's canned salmon was an inexpensive staple in the diet of many Americans. The more economical gold mines could operate even before the price of gold was increased to thirty-five dollars per ounce. Federal largesse, always a factor in Alaska's economy, expanded from its Alaska Railroad base as the depression-era economic-stimulant programs kicked into gear. Alaska received Works Progress Administration (WPA) funds. The federal-owned RFC subsidized some Alaskan mining projects. In the Matanuska Valley, the federal government embarked on the Colony Program, which transported and then subsidized farm families who moved from the northern Midwest to Alaska.[1]

Alaskans also enjoyed a healthier and happier lifestyle than many other Americans. They did not stand in line for apples or soup. Most Alaskan natives and many other residents throughout the territory still lived a semisubsistence life. They could pick berries, catch salmon, kill a caribou, or buy a locally produced reindeer steak. Many rural and city residents harvested rhubarb or giant cabbage and other produce from abundant, if short-lived, gardens.

The inhabitants of the great land were also spared most of the liquor- or strike-induced violence of the period. Anchorage's homegrown bootleggers were probably more of a danger to themselves than to anyone else if they consumed too much of their own product. Alaska's port cities of Juneau, Ketchikan, Seward, and Cordova were inconvenienced when Harry Bridges's communist-infiltrated longshoremen's union struck, but they were spared the type of violence that occurred during strikes in Seattle and San Francisco.

Mining matters and metal prices for both gold and silver often made the news. The same day that the *Anchorage Daily Times* noted the expenditures for the Matanuska Colony, a new world silver price, equal to that of the U.S. Treasury, was announced. Gold was consistently newsworthy, and Dunkle was often in the headlines of newspapers in Anchorage and Fairbanks. From 1931 until 1941, Dunkle apparently never kept fewer than one hundred men working at his mines. At the high point in the mid-thirties, the total number of employees on Dunkle mining projects must have exceeded two hundred; seventy-five more worked for Star Air Lines. As Peter Bagoy Sr. explained, "In the thirties if you weren't working for Dunkle you weren't working."[2] Pete knew this was an exaggeration because he himself had worked for the Alaska Road Commission, which was an even larger employer. The Alaska Railroad and Fairbanks Exploration Company, a subsidiary of U.S. Smelting, Refining, and Mining Exploration Company (USSRM), employed more men than Dunkle did. Many hundreds of people were employed at the canneries and in the fishing industry. Few single individuals, however, kept as many people working in rural mainland Alaska as Dunkle did during the 1930s. The jobs also had quality. Dunkle had learned during his apprenticeship with Kennecott Copper Corporation that well-housed and well-fed employees were also productive employees. The miners' wages of about eight dollars per day now sound minuscule, but Dunkle's pay rates were more than competitive for that time.

Dunkle's direct contributions through payroll and purchases of supplies and equipment were augmented by economic multipliers. Trade accounts with heavy-equipment dealer Carrington and Jones, Anchorage grocery wholesaler W. J. Boudreau, Northern Commercial Company, A. A. Shonbeck, and many others supported other jobs and spread wealth in Anchorage and back through Seattle to conterminous America. Some Dunkle dollars moved outside Alaska via the purchase of airplanes, automobiles, bulldozers, and draglines in New York, Detroit, Peoria, and Milwaukee. Other Dunkle contributions to the economy were made through his investments in local stocks and his loans. Dunkle loaned H. I. Staser ten thousand dollars in start-up money for a mine north of Girdwood, Alaska. He told Staser, "I never expect to see [the money] again. Your area is geologically unsound."[3] Dunkle was proven right when Staser's vein was cut off by a fault, but not until after Staser had mined four thousand ounces of gold and had repaid Dunkle within the first month of the operation.

As in the Kennecott years, Dunkle looked at many more prospects than he attempted to develop. During the 1930s, Dunkle's area of prospecting activity extended from the Brooks Range to the panhandle of southeastern Alaska. He extended his range through the observations of other engineers, including a young James Stanford and an experienced Bert Nieding, who was formerly a general manager at Kennecott Copper Corporation. In 1935 Dunkle looked at grassroots properties in the Goodpaster region east of Fairbanks, where a 1934 gold discovery had prompted a minor gold rush. He was in the area soon after the discovery and covered much of the ground with prospectors Bill McConn and Carl Tweiten. Dunkle believed that he had firm options on some of the ground through an Alaskan character named Jack McCord. McCord, however, either reneged or lacked control over the options, and Fairbanks Exploration Company moved in to undertake the first underground development. Perhaps because of its remoteness, the district never prospered. (In 1995, only a few miles downstream, a tenacious geologist drilled a hole that intersected a rich gold vein several hundred feet below the surface. By the fall of 2000, drilling had blocked out more than five million ounces of high-grade lode gold at the Pogo prospect.) Dunkle also looked very hard at the Oracle mine northwest of Moose Pass, which was part of a widespread gold belt in the Chugach Mountains. He spent more than twenty-five thousand dollars on the prospect in 1933 and 1934, but abandoned it shortly afterward.[4]

One district, Flat (Iditarod), was not abandoned so quickly. Dunkle decided that the district in southwestern Alaska was worth a major effort because of its size and richness. Since 1909 Flat has produced more than 1.5 million ounces of placer gold, following only Fairbanks and Nome in Alaskan placer productivity. Although a few other placer fields were discovered later, a discovery on Otter Creek by Johnny Beaton and W. A. Dikeman on Christmas Day in 1908 led to the last great gold stampede in Alaska.[5] The discovery on Otter Creek was followed by an even richer one on the Marietta claim near the head of Flat Creek. Almost all of the creeks radiating from Chicken Mountain above the Marietta claim were gold bearing. By the mid-1920s two other prospects had been identified at Flat. One was a large but low-grade deep placer deposit on Willow Creek southwest of Chicken Mountain. The other was an incredibly rich lode deposit called the Golden Horn. Dunkle developed both properties.

The Golden Horn lode, discovered and named by Rasmus Nielsen in the early 1920s, seemed rich enough to be mined. In 1925 Nielsen and mining partner Jerome Warren shipped a little more than eleven tons of ore to the smelter at Tacoma. The shipment contained gold and silver equivalent in value to more than thirty-four ounces of gold per ton. In 1926 Justus Johnson and Paddy Savage acquired the property on a lien. Johnson and Savage, with the assistance of Warren, shipped another eleven tons, which netted a return equivalent of more than ten ounces of gold per ton after deducting $32.50 per ton just to haul the ore as far as Iditarod, about ten miles from Flat. Other shipments to Tacoma contained more than fifteen ounces of gold per ton. The mine remained small, however, partly because of groundwater. Even with the financial assistance of Minnie Warren, Jerome Warren's wife and a sometime prostitute at Flat, Johnson and Savage had difficulty developing the mine. They needed the help that Dunkle could provide.[6]

Dunkle and Jim Stanford arrived at Flat in May 1934 to map and sample the Golden Horn.[7] Their assays confirmed that the ore was rich but suggested that the vein was narrow. It averaged about three ounces of gold per ton over a one-foot width, which was sufficient to carry a mining width of about five feet. Locally the vein was very rich, as much as twenty ounces of gold per ton. On the strength of the results and the geology of the district, which suggested that erosion from the Golden Horn had fed a large part of the highly productive Otter Creek placer, Dunkle optioned the ground from Johnson and his partners.

Dunkle was confident enough to option the property without consulting Pardners Mines or L. C. Thomson, but he offered them a controlling interest in the mine. He also wanted Bert Nieding to oversee the project:

> I would like to have you come up and take charge of the work but cannot offer you a very large salary for I shall only have a small crew

> to start with, probably about eight or ten, enough to sink a couple of hundred feet and do a thousand feet or so of drifting. I could pay you $250 per month and expenses. . . . I have offered Thomson and Baragwanath of this company [Willow Creek Mines, Inc.] the chance to share with me in the development and do not know if they will come in. . . . If they do not come in and I have to do it all myself I shall also give as a salary bonus 10% of the operating profit.[8]

In mid-July 1934, Dunkle sent Baragwanath the assay plan of underground workings and the results of surface sampling. He did not equivocate on the width of the vein:

> You will note that the vein is very narrow. However, it is very high grade in spots, some of the preliminary grab samples I took going as high as $700 [twenty ounces of gold per ton at $35.00 per ounce]. . . . I consider the chances of making a good small mine out of it better than fifty percent favorable. You will note the remarkable persistence of at least moderate values at all points.
>
> I have ordered a hoist, drills, steel, compressor, lumber for camp and shaft and am prepared to sink 200 feet and drift 1000 feet and confidently expect to have $500,000 gross in sight of at least $50 value when that much work is completed. I would like to see you come into it. Enclosed are copies of the purchase agreement.[9]

With a formal submittal to Baragwanath out of the way, Dunkle again wrote to Nieding to make sure he understood the deal, including the bonus. Nieding accepted, went to Flat, and began to construct a mine (fig. 29) with project funds funneled through Dunkle's longtime banker, E. B. Kluckhohn at the Seattle First National Bank. In late October Dunkle telegraphed Kluckhohn: "Please forward following telegram to Baragwanath quote estimate fifty thousand dollars will carry Golden Horn to next summer . . . believe one hundred and fifty thousand more will put it on a profitable basis stop I have already advanced more than thirty thousand and would appreciate refund of your portion."[10]

On 28 October Baragwanath advised via telegram, "Thomson and I have decided take two-thirds participation Golden Horn for Willow Creek. Understood that if development on next level unfavorable Willow Creek can withdraw at its discretion and need not be obligated for its share further development. . . . Have telegraphed Kluckhohn pay you twenty thousand representing two-thirds amount advanced by you."[11]

Results from the project appeared to be favorable into early 1935, although Nieding complained to Dunkle about the high prices that he

Figure 29. Golden Horn mine, Flat district, looking north across Otter Creek, 1986. The head frame and hoist house were built in 1934. The dumps to the right of the hoist house stockpiled "low-grade" ore that contained more than an ounce of gold per ton. Courtesy, John Miscovich, Orange, Calif.

had to pay for mine timber. Nieding set up an assay lab at Flat, and Dunkle reported that check assays, done at the Lucky Shot mine, showed rich ore. On the basis of high assays received in late 1934, Dunkle advised Nieding to crosscut a wider interval around the vein: "Apparently there is some kind of very high grade mineral, either a telluride, or else arsenopyrite of high gold content. If it is telluride it can form on the joint faces of the granite and make some wide ore bodies. It is a point we should determine as soon as possible." Nieding had anticipated Dunkle's request with a crosscutting program that mainly proved that he was following the vein and that it was narrow. Later in the spring, he began to develop the mine for production, driving raises and other development headings. It was not a simple task, but Nieding had the necessary experience to stay in ore:

> We finished the raise on the second of this month. . . . The ore changed its dip every few feet, and was faulted twice. . . . I followed

> the ore all the way and holed out in the first level, just north of where Justus had started that last little chute raise. I would say that the ore will average about six inches in thickness, and will assay from thirty to nine hundred dollars per ton. . . . [T]he Wall rock invariably carries about five dollars a ton, some samples went as high as twenty and some as low as seventy cents.[12]

Nieding also advised that the ground conditions were not pleasant: "The ground requires timber for it will slough without warning, although it sounds solid. Water following thru the seams adds to the difficulty." In the same letter, Nieding noted that he had read that Dunkle's plane was on its way to Flat and expressed, "I will be dam glad to see you and get your ideas on the situation here."[13]

Nieding began to mine and ship ore in June 1935. He hit a thin streak of ore in the north workings that assayed about fourteen hundred dollars per ton (forty ounces of gold per ton), but in general, he told Dunkle, "things look like hell." Dunkle was on the way to Flat by 13 June. He wanted Nieding to start a stope in an area that sampled well, even if it did not look like ore, and to stop driving prospect openings and just mine. Dunkle was flying the little Travel Air with limited fuel capacity; he told Nieding, "I shall have to duck around a little."[14]

At the Golden Horn, Dunkle found ore as rich as he had anticipated but in a vein as narrow as he had feared. The vein was complexly faulted and perhaps even cut off by a fault to the south. But there was rich ore to the bottom of the mine. Ore in a sump below the lowest level contained seventeen ounces of gold per ton. Nieding also worked to upgrade the ore as much as possible before it was hoisted from the mine. He retained an old Serbian miner, Popovich, who did nothing but hand sort the material in the stopes. Under Nieding's management, the Golden Horn mine shipped about 250 tons of high-grade ore to the smelter at Tacoma. Much of it assayed about six ounces of gold per ton. More than one thousand tons of rock rich in arsenopyrite and scheelite (an ore mineral of tungsten) were stockpiled for later processing. This material contained more than an ounce of gold per ton but was not rich enough to net a profit after freight and smelting charges were deducted. Also working against profitability was a gross royalty of 25 percent paid to Justus Johnson and his partners. The rate was high but not unheard of for a rich vein. At remote Flat, however, it was a serious impediment to successful development of any mine. Justus Johnson and his co-owners apparently did not wish to reduce the rate, perhaps because they now had a developed mine and a valuable stockpile.[15]

Golden Horn had gross receipts of about $25,400 for 1935. Labor and supplies, however, had cost more than $37,000, yielding a net loss of almost $12,000. Johnson and the other mine owners received $7,452 in royalties. Willow Creek Mines, L. C. Thomson, and Harold E. Talbott took about a $48,000 loss over the life of the venture.[16] The Golden Horn mine was returned to its owners in 1936, but a property that an experienced Bert Nieding could operate with pumps working at least one shift per day could not be operated on a shoestring. The owners were only able to high grade the upper part of the workings, as they had done in the years before Dunkle's operation.

Although the Golden Horn venture was not profitable for Dunkle and Willow Creek Mines, it was a significant boost to the local and general economy. Between 23 July 1934 and 31 December 1935, Dunkle, for his one-third interest, sent about $24,400 to Carrington and Jones for building supplies and mining equipment. An even $20,000, doubled by Dunkle's partners, went to Nieding for payroll. A. A. Shonbeck received $2,074 and, through his Standard Oil pocket, probably another $2,944. Dunkle paid the Alaska Road Commission $2,000, probably for repair of the road to Iditarod. Harry Donnelley's Miners and Merchants Bank in Iditarod took $5,000 for its cut. (Donnelley, Flat's millionaire, made money from this transaction and at times from others when no one else did.)[17]

Dunkle's other project at Flat, the lower Willow Creek placer deposit, proved to be too difficult to mine. Now, more than sixty years later, old-timers at Flat still debate the causes of the failure.[18] It was not simply the result of a lack of information. The property was drilled by its owner, Frank G. Manley, in the 1920s with marginal results at a gold price of twenty dollars per ounce. After the price of gold was raised, the property looked better. As stated by the *Fairbanks Daily News-Miner,* the "increased price of gold has greatly enhanced the value of the property and increased the workable area of the pay streak on the property."[19] Dunkle did not take the enhanced results at face value. To confirm previous drilling, Dunkle hired rotund Ben Bromberg, who was almost always called by his nickname, "Jew Ben." Bromberg was one of the best placer gold drillers in the territory. He redrilled the three-and-one-half-mile-long placer deposit. His results, with values of twenty-five to thirty cents per *bedrock foot* (see glossary), were still at the lower end of the economic limit. A large-scale placer mine could make a profit if everything worked.

Perhaps Dunkle's chosen manager, Peter T. Jensen, contributed to the ultimate failure of the project. There are two views of Jensen as a miner. One was reported in the *Anchorage Daily Times* and repeated a few days later in the *Fairbanks Daily News-Miner.* The newspapers called Jensen a "pioneer Alaskan placer operator" and noted, "Mr. Jensen has been mining in Alaska and the Yukon since he first stampeded to the Klondike in the rush days over thirty years ago."[20] The other view of Jensen was that he was an excellent cook and later a successful restaurateur, who may have been around northern placer mines for thirty years but mainly near the cookhouse. The other necessary part of a large-scale operation is the equipment. Like the manager, it is still a subject of some contention at Flat. It was big and costly and should have done the job. The largest single component of the mining fleet was a four-cubic-yard P & H dragline that weighed 180 tons. There were two TD-40 International tractor bulldozers, a ten-inch diesel-driven pump, a two-and-a-half-ton White diesel truck, and a mobile elevated sluice box. All the equipment was new, and all or most of it was assembled through Carrington and Jones of Seattle and Alaska.

In the project's first year (1934), before the arrival of the big dragline, Jensen hydraulically stripped a half million yards of barren overburden (*muck* to the miners) to allow solar thawing of the pay gravel to begin. The following March, he waited until breakup to move the dragline the last leg up the Iditarod River and set it in place. The dragline was already a year late, as Dunkle wrote Baragwanath in the summer of 1934: "Have given up all hopes of erecting the dragline this year but hope to have it up early next spring. The prospecting results on the ground are continuing very satisfactory."[21] When the big dragline finally arrived, it was erected at Flat, which was a considerable undertaking (fig. 30). In 1936 there were still problems, as Dunkle again reported to Baragwanath:

> Pete is having more trouble getting started than he thought he would on account of the large amount of work required to get in his bedrock drain down Sourdough Creek. He expected to be sluicing by the middle of July but it will be about the middle of August before he is ready. . . . When he does get started sluicing, he will have good gravel but it is bound to have delays and I do not see how he can hope to take out more than $50,000 gross this year, and little, if any, of that would be available for distribution.[22]

The ultimate test of mining is the ground itself. The ground in lower Willow Creek was deep, frozen, and as low grade as drilling had indicated. The operation was not profitable. Dunkle invested at least fifty-five thousand dollars in the project. Baragwanath and Thomson invested

Figure 30. Erecting (left) *and operating* (right) *four-cubic-yard P & H dragline at Dunkle's Iditarod Mining Company property, lower Willow Creek, Flat district, 1935. Courtesy, Dunkle family.*

even more. Dunkle and Pardners Mines wrote off the project in December 1937.[23]

As at the Golden Horn mine, the positive results of Dunkle's efforts at Willow Creek were to the larger economy, not to his Iditarod Mining Company partners. Glenn Carrington and his associate Arthur Erickson of the Carrington and Jones company told John Fullerton, who was later a successful operator in Willow Creek, that the early-depression purchase of the big dragline from P & H in Milwaukee helped that firm through some rough times, before the depression-era mining orders began to pour in. It certainly helped the Carrington and Jones firm. Later all the equipment shipped to Willow Creek was of value to others: first of all, to Flat's Harry Donnelley, who picked up the machines for a song. Fritz Awe, operating on Chicken Mountain at Flat, needed the big dragline for his operation. Donnelley sold it to him, and Awe prospered. Toivo Rosander and Joe Degnan needed the International tractors for their operation at Ophir in the Innoko district. John Miscovich "walked" the tractors into Flat, and Rosander and Degnan "walked" them over the winter trail to Ophir. The ten-inch pump went to Alec Matheson, who was mining on lower Flat Creek. After Fritz Awe completed his operation on Chicken Mountain, Richard and John Fullerton put the dragline to work. It was in use in Happy Creek in 1968 and was still operating in Willow Creek in 1976.

Dunkle had better luck with placer development projects in two other districts, although he elected to let other operators proceed with them. The districts were Circle, almost on the Yukon River north of Fairbanks, and Yentna, which was closer to Dunkle's home base at Anchorage. At Circle, Dunkle and Pardners Mines drilled Half Dollar Creek and looked favorably at mining nearby Portage Creek. In the Yentna district, Dunkle supported placer mines in Bear and Dutch Creeks through investments in Fern L. Wagner, whom Dunkle trusted as a competent operator. Dunkle and Pardners also made early investments in the Fairview placer mine at Collinsville in the southern part of the Yentna district.[24] The investments in the Yentna placer district appear to have been on the plus side. Wagner repaid his loans. The Fairview mine later proved to be a nice operation but was probably too small to appeal to Dunkle and Pardners Mines.

One of Dunkle's favorite projects in the mid-1930s was Bill Shannon's lode prospect at Slippery Creek in Mt. McKinley National Park. Dunkle wrote to all of his usual correspondents about the deposit, which he described as an igneous dike mineralized for at least three miles. The dike dipped about seventy degrees. Both the hanging wall and the footwall were mineralized over widths ranging from five or six feet to as much as twenty-two feet. The arithmetical average of Dunkle's samples approached 0.3 ounce of gold per ton, and none were less than 0.15 ounce.[25]

Anaconda Company also liked the Slippery Creek project, but a company geologist had reservations about the metallurgy. Francis Cameron wrote to Murl H. Gidel, the assistant to chief geologist Reno Sales, on 27 July 1937: "The Shannon sample when it arrives should be studied with care to determine mineral associations and tests made to determine its flotability and to determine if it will be possible to cyanide. This ore in appearance is similar to Whitewater ore and contains both antimony & arsenic. It looks difficult to treat, but I wish really careful tests made as there are distinct possibilities of a very large tonnage." Dunkle and Pardners Mines pursued the project diligently in 1936 and 1937 and invested at least fifty thousand dollars.[26]

Dunkle had another reservation about the Shannon prospect, which he shared with Baragwanath: "I do not like the feature that it is situated in the heart of the Park [Mt. McKinley National Park], but at present the mining laws are the same in the Park as they are out of it, and will probably remain that way unless some of our political friends get the idea of grabbing off something good by means of the Park regulations."[27] (Mt. McKinley National Park was open for mine location from its formation in 1917 until the passage of the Alaska National Interest Lands legislation in 1980.)

The first mining and assay tests from Slippery Creek tended to confirm the size potential of the property, but Anaconda's metallurgical tests of fresh (not surface oxidized) material found that the ore was highly refractory. Less than 10 percent of the gold in the bulk sample that Cameron shipped to Butte was available for leaching.[28]

Geologically the Slippery Creek prospect remains an interesting deposit deep within Denali National Park and Preserve. Dunkle and Pardners Mines wrote it off at the end of 1937 because of the results of Anaconda's metallurgical testing and because underground development indicated an ore grade even lower than that of the surface samples. During 1936 and 1937, however, the Slippery Creek project kept another ten or so people working and distributed dollars to the Alaska Railroad; McKinley Park Transportation Company; the prospect owner, Bill Shannon; and the usual Dunkle trading partners: Star Air Service, W. J. Boudreau, Arthur A. Shonbeck, and Carrington and Jones. Although Dunkle wrote off Slippery Creek, he remained interested in the geologic potential of the central Alaska Range, a belt of rocks that had intrigued him since 1915.[29]

10

The Golden Zone

DUNKLE'S FIRST TRIP INTO THE CENTRAL ALASKA RANGE WAS IN 1915, when he appraised the Broad Pass region and its chief prospect, Golden Zone, for the Guggenheims. He first saw the north flank of the range in 1923, when he explored the zinc-rich deposits at Mt. Eielson and reconnoitered as far west along the range as Bill Shannon's prospect at Slippery Creek. Dunkle returned to the north flank in 1928 on an examination of the Little Annie vein of Joe and Fannie Quigley in the Kantishna district. Periodically, at least in 1923, 1928, and 1934, he revisited Broad Pass to gauge its progress.

After two decades of scouting and developing properties in Alaska, the Pacific Northwest states, British Columbia, and Africa, Dunkle remained intrigued with the potential of five properties in the central Alaska Range: lode deposits at Little Annie, Liberty Bell, Slippery Creek, and Golden Zone and a placer deposit in Caribou Creek. By 1937 all but Golden Zone and Caribou Creek had been eliminated from Dunkle's priority list. The Liberty Bell property was not available, at least on terms that Dunkle liked. Work by Ira B. Joralemon on the Little Annie prospect indicated that the ore was lost, possibly irretrievably, at a fault. Dunkle's and Anaconda's work at Slippery Creek showed that the ore was refractory and could not be worked at its remote location.[1]

Neither of the remaining properties—Golden Zone and Caribou Creek—could have been developed when Dunkle came into the country in 1915, but there had been a host of tangible and intangible changes since then that benefited both properties. The first major barrier to development fell with the completion of the Alaska Railroad in 1921. A ribbon of steel traversed the region. Because of the railroad, Dunkle could be in Fairbanks or Anchorage in two days or less from any point in the range, and conversely, heavy freight, coal, and fuel oil could be landed at almost any point along the line. Telegraphic communication was available from any of the numerous maintenance stations; telephone service soon followed. For passengers and light freight, air transport was much faster and, for a successful gold miner, not too expensive. By 1928 skilled pilots, such as Joe Crosson, were landing on Quigley's rough airfield at Kantishna near the Caribou Creek and Little Annie prospects. By 1938 there were airfields at many other sites, including the Stampede mine, Savage River, and West Fork of the Chulitna River just below the Golden Zone. Especially after the depression began, federal funds were available for the construction of resource roads and trails from the railroad to the mines themselves.

Dunkle believed that coal and electrical energy could also come from the region. There were large coal deposits north of McKinley Station and possibly to the south at Broad Pass. The innovative engineer had been intrigued with the potential of hydroelectric power at least since his work for Alaska Steamship Company on power sites in southeastern Alaska. He was certain that he could tap fast-flowing streams for power and believed that he might even be able to extract energy from low-gradient rivers, such as the West Fork of the Chulitna. Water was also important in its own right. Although water crossings were a hazard, water was necessary in both lode and placer mining, and rivers formed natural highways that supplemented the lone rail line.

Less tangible factors contributed to Dunkle's plans. Anchorage and Fairbanks were small cities. Civic leaders who may have been prospectors themselves a few years before lent their names to mining ventures, invested in mines, and supplied the needs of the miners. One of Dunkle's friends, Glenn Carrington, owned the dealership for International Harvester equipment. Another friend, A. A. Shonbeck, was the dealer for Caterpillar equipment, Ford cars and trucks, Standard Oil products, and Giant explosives. Anchorage- or Fairbanks-based attorneys could draft corporate agreements. On a personal level, Billie Rimer (soon to be Dunkle) jumped into the social life of Anchorage and quickly made friends with the wives of civic leaders: Mrs. Warren Cuddy, Mrs. Robert Atwood, Mrs. Frank Reed, and the Walkowskis and Romigs. Billie's contributions were not only social. She was a trained scientist and an inventor in her own right, who was capable of helping Dunkle with technical problems.[2]

After 1934 gold could be sold to the U.S. Mint for $35.00 per ounce instead of $20.67, and there was nothing to hold Dunkle back. With the assistance of Pardners Mines, he initially concentrated on the Golden Zone project, where he took advantage of improvements made and geologic data acquired since 1915. Geologist C. P. Ross of the USGS remapped the district in 1931 and reported that work had begun to extend the old Guggenheim tunnel (the one-hundred-foot level, or 100 level) another 125 feet. The results of the tunnel advance were sufficiently positive that Guggenheim-related ASARCO optioned the property in early 1932 and began a new exploration project. ASARCO's miners dropped down the hill a hundred feet vertically and collared in a new adit: the 200 level. Initially the rock was nearly barren granite. After driving two hundred feet, miners crossed a sharp contact and hit mineralized *breccia,* fragments of granite cemented with gold-bearing quartz and sulfide minerals. The drive continued in the breccia for the next 146 feet. The breccia was appreciably mineralized from 250 to about 300 feet, assaying about eight dollars per ton in gold. The grade then dropped off; at 346 feet into the mountain, ASARCO quit.[3]

Dunkle disagreed with ASARCO's assessment, especially after he sampled the extension of the 100 level. Ore in the extension locally contained more than one ounce of gold per ton and averaged about 0.4 ounce. Dunkle was virtually certain that ASARCO had not driven the 200 level far enough to test the downward limit of the better ore. He was also certain that other conditions had improved since he was at the Golden Zone prospect in 1915. Dunkle told the *Fairbanks Daily News-Miner* that the combination of an increased gold price and more efficient mining machinery made the property much more attractive than when he first examined it and that he did not agree with the ASARCO appraisal: "Mr. Dunkle on the other hand is of the opinion that the property is worth testing."[4]

Dunkle and Pardners Mines optioned the Golden Zone property from Lon Wells in August 1935. By March 1936 employees had been hired and supplies and equipment began to move north via Alaska Transfer Company and the Alaska Railroad. On 10 April Dunkle made a ten-thousand-dollar investment in the Golden Zone. Accelerated activities after that can be tracked through purchases from Boudreau and Ship Creek Meat for groceries and through Shonbeck and Alaska Construction and Lumber for other supplies. Funds were sufficient to purchase a sawmill for five hundred dollars, a Pelton wheel for hydroelectric power generation, and a new International bulldozer from Carrington and Jones.[5]

As at the Lucky Shot mine, the Golden Zone project looked sufficiently large and complex that Jack Baragwanath and Harold Talbott of Pardners Mines wanted an independent appraisal before they committed to participate in the project. Again Talbott sent Dunkle's classmate Henry Carlisle to make the examination. Carlisle's main concern was metallurgical: the recovery of gold from an ore rich in arsenic minerals.

Golden Zone ore was even more arsenical than that of the Shannon claims, where Anaconda found that most of the gold could not be recovered economically. Arsenic-rich gold ores, however, differ. In some ore, the gold is tied up in the atomic structure of the arsenic minerals and is difficult to recover. In other arsenic-rich ore, the gold was formed as minute crystals in the sulfide and arsenic minerals and can be liberated through fine grinding. Dunkle shipped Carlisle's samples to San Francisco for testing. The tests showed that at least 80 percent of the gold in Golden Zone ore was "free," unbound to the atomic structure, and amenable to cyanide leaching.[6] Carlisle endorsed the Golden Zone prospect, and Pardners Mines backed Dunkle at the property.

Construction at the Golden Zone mine began in the late spring of 1936. By late July, Dunkle reported to Baragwanath:

> The pipe line and power plant [are] installed, ditches and dams in, etc. The water wheel and generator [are] installed but the switch-board work is not completed. The pole line is about three-fourths completed. We have built a fair tractor road from the Riverside to the Golden Zone, about three miles, and a rise of fifteen hundred feet. . . . The bunkhouse at the mine is nearly completed and will make very comfortable quarters for sixteen or twenty men. The excavation is complete for the compressor building and the blacksmith shop. . . . We are widening and retimbering the lower [ASARCO] tunnel, which was driven with a wheel-barrow and is too narrow for a car. The upper tunnel will not require so much work, but, of course, we must lay track in both of them. We have got a lot done, all right, but there was a lot to do, and it has taken a lot of time.[7]

Mining development continued throughout the year. The lower tunnel (200 level) was advanced, and the miners hit heavy sulfide breccia at five hundred feet. The breccia locally assayed more than six ounces of gold per ton and averaged about one-third of an ounce. The ore was exactly where Dunkle predicted it would be. Turning south, the miners stayed in ore for 250 feet. Raises driven upward in the ore showed that good ore was also continuous from the 200 (ASARCO) level to the surface. The next step was to determine if the ore went deeper. The miners cut several diamond-drill stations and began to drill horizontally and downward. Several of the holes hit ore-grade material. Together the pattern of ore intercepts indicated that the main mineral zone extended downward at least three hundred more feet, for a total distance of five hundred feet from the outcrop of the ore body. Grades were good, but more exploration was needed before the decision to mine could be made.

The project also needed better access. The company had built three miles of road in 1936, but the railroad was still ten miles away and separated from the mine by the wide West Fork of the Chulitna River as well as narrower, but still troublesome, stream crossings at the Middle Fork of the Chulitna, the Bull River, and Costello Creek. The Territory of Alaska and the federal government helped solve the access problem. In the report for the fiscal year ending 30 June 1936, the Alaska Road Commission recommended an appropriation of $1,580,000, in addition to regular funds, for the entire territory. Recommended projects included "construction of twelve miles of road from the Alaska Railroad serving a developing mining area." In 1937 funding was augmented by $110,000 from the Emergency Relief Agency and $450,000 from the WPA. Of these funds, $350,000 was to be spent on twelve territorial road projects, including "Colorado Station on the Alaska Railroad."

At that time, the Alaska Road Commission estimated that it could build in Alaska's interior a gravel road that would be adequate for summer use for $8,000 per mile and could maintain it for $350 per mile. Dunkle's road project came in under budget. In 1937 the commission spent $60,182 on the Colorado Station project (Project No. 93A) to build nine miles of road and more than three thousand feet of timber trestle bridge.[8] Peter Bagoy Sr. was dirt foreman on the project; he supervised the line cutters and operators. The other boss, the bridge foreman, brought in a few iron parts and then fabricated and spliced local timber together to fashion a pile driver. Bagoy remembered that construction started from a camp at Colorado Station. After the workers completed a short airfield and extended the road to Bull River, the camp was moved to the river. It was moved one more time to Colorado Creek, which was the base for building a bridge across the West Fork of the Chulitna River. Construction began about 1 July 1937 and was nearly completed by Thanksgiving. Just before the road builders left for home, they invited the Golden Zone crew for Thanksgiving dinner.[9]

By the end of the 1936 season, the Golden Zone mine was looking big. Although the grade of most of the ore ranged from 0.1 to 0.3 ounce of gold per ton—in contrast to the Lucky Shot's 2.5-ounce material—the deposit was more compact. At the Golden Zone, Dunkle could see more gold in a massive body contained within one mining claim than he had recovered at the Lucky Shot mine from an ore body nearly one mile long but only a few feet thick. It was also apparent, however, that extraction of the gold would be more difficult than at the Lucky Shot mine and that development could be costly. Under those circumstances, Dunkle and Pardners looked for a major company to develop the Golden Zone mine. Anaconda Company was a possibility, although not all of Dunkle's friends and associates were enamored by that prospect. Probably reflecting decades-long competition between the companies, an old Kennecott and African friend, J. F. Bowes, wrote Dunkle from England: "I am sorry to hear that you are contemplating negotiations with Anaconda. I should have thought you were too respectable to consort with such scoundrels. I don't know what Stannard would think of it."[10] Kennecott was probably Dunkle's first choice, but Stannard knew about the project, and Guggenheim-affiliated ASARCO had abandoned it in 1934. It would have been a very hard sell to bring any Guggenheim-related corporation to the property in 1937. Moreover, Dunkle did not share Bowes's view of Anaconda. Harry Townsend, Anaconda's Alaskan geologist, was one of Dunkle's best friends, and Dunkle respected Anaconda's geological ability.

Anaconda optioned the Golden Zone mine in midsummer 1937, and the company sent Townsend and Francis Cameron to the property to make a confirmatory field survey. Initially Anaconda was enthused by the project. As at the Shannon property in Mt. McKinley National Park, Anaconda began work at the Golden Zone with its own metallurgical study. In late July 1937, Cameron shipped two bulk samples of Golden Zone ore to Butte for testing, noting, "I have options on both the Shannon & Golden Zone and I feel that both warrant very careful consideration. I am returning to Seattle as soon as I can get going on these, and I would like to see Reno [Sales, Anaconda chief geologist] regarding them at the first opportunity."[11] In late September, after mapping and sampling the underground workings and examining the results of Dunkle's first thirteen drill holes, Cameron was still enthused. As to the possibility of ore, Cameron reported to Reno Sales, "Present development has not blocked out a proven ore tonnage but has indicated possibilities of substantial tonnages if the horizontal sections, previously referred to, continue to depth. There is no reason to expect that they will not continue to, say, a depth of 1000 feet or more."[12]

Problems emerged, however, when Anaconda began its own confirmatory drill holes. Diamond drills today usually recover 90 percent or more of the rock. Modern core samples are large in diameter, from two to more than three inches across. The diamond-faced cylindrical bits that recover the core are designed to drill material of varying hardness. Different hole patterns are used in the face of the bit to deliver water and mud lubricants to the cutting surface of the bit, depending on the material being drilled. The diamond drills used at the Golden Zone mine were primitive in comparison. Cores were only about one inch in diameter, and core recovery was poor, especially in the breccia, where the hardness of the material changed abruptly as the bit passed from granite to quartz to sulfides and even to clay. The breccia can easily be drilled today. The modern procedure with good core recovery is to split or saw the core lengthwise, assay one of the splits, and save the second split in a core library. In 1937 poor recovery was accepted as a fact, so the drill cuttings contained in the water that splashed back from the hole (called *sludges*) were saved and assayed. Rather complex equations were used to calculate ore grade from the combined partial-core and sludge assays. This was the method that Anaconda used in core holes drilled on the 200 level. The core recoveries were poor, and assays of the core itself generally indicated very low gold values. Assays of the sludge samples were much higher; in fact, they indicated mineable ore.

Dunkle's experience with drilling the breccia led him to trust the sludge assays. Cameron conservatively favored the assays from the core. As in many similar situations, company geologist Cameron, who had returned to Seattle, leaving Townsend in charge at the mine, turned down the Golden Zone venture:

> I received core samples from Colorado and Harry's data on drill results to February 9, on the 18th. A study of these data makes the picture look far from encouraging. It looks as though we are not getting any place with our drilling and cannot under existing conditions. . . . Pat Lynch [Anaconda's drill expert] . . . is of the opinion that this can only be done from the surface with an independently powered drill. It would mean starting a 3-inch hole and casing all the way. I feel that our drilling results are anything but satisfactory and due to caving of the holes and poor core recoveries any estimate of grade is somewhat of a guess. To continue drilling . . . would, I think, be foolish and a waste of money.
>
> It thus appears to me to be foolish to continue in the face of conditions to date, and as I was unwilling to recommend making payments called for under our agreement with you I wired to this effect as soon as I could.[13]

On the same day that Cameron wired Dunkle, Dunkle wrote to Cameron that he was encouraged by the drilling results and the high assays of the sludge.[14] The letters crossed. On 11 March Cameron wrote again, reiterating his previous conclusions:

> I note from your letter that you do not agree with me as to the proper weights to assign Golden Zone sludge and core samples in combining same to get an approximate average ore grade, and from

> your interpretation of results the average grade is much higher than I have allowed.
>
> That there are possibilities of considerable error regardless of the method of combination used I will readily admit since recoveries of both core and sludge were not satisfactory. But that I am entirely in error in this instance . . . I do not admit.[15]

Although Dunkle argued by mail with Cameron, he recognized that decisions like this are not usually reversible. Cameron continued in the 11 March letter: "I note that you have taken the matter of continuing development up with the F. E. Company [Fairbanks Exploration Company, a subsidiary of USSRM and operator of the dredge mines at Fairbanks]. They may see things differently than we do and I hope for your sake that they do. In a way the results of holes 16 and 17 make the picture a bit more attractive if they think they can make low enough costs and can keep the capital investment down." Dunkle, however, could not reach an agreement with F. E. Company. Another letter from Dunkle to Cameron suggests that the Golden Zone owners turned down F. E. Company's proposal because it was only a commitment to do further testing, not to develop the mine.[16]

The effect of Anaconda's withdrawal could have been fatal to the project, but Dunkle was still financially healthy, and he had friends and business associates who would join in a development project. Many citizens of Anchorage were sufficiently prosperous that they could afford a well-conceived mining venture, the type of project that they expected from Dunkle.

The commercial vehicle chosen by Dunkle and a few close associates was a public stock corporation, Golden Zone Mine, Inc. (GZM), headquartered in Alaska. Dunkle signed over the lease on the property to the corporation on 14 May. Two days earlier, the company's promoters had placed their signatures on a stock subscription form that also defined corporate responsibility and named the officers: Dunkle, president; Alexander H. McDonald, vice president; and Edgar R. Tarwater, treasurer. The subscription form itself was a series of blanks following the name of its first subscriber, Warren N. Cuddy. Under the terms of the corporate structure, Dunkle kept the rights to the majority of shares (55 percent). Fifteen percent was reserved for mine laborers, and another 15 percent was reserved for Lon Wells, the surviving discoverer of the prospect. Operating capital was to be raised by the sale of one million shares of common stock at forty cents per share (fig. 31).[17]

The officers of the corporation were pioneer Alaskans who were well regarded in the Alaskan business world. The corporation's vice presi-

INCORPORATED UNDER THE LAWS OF THE TERRITORY OF ALASKA

No 314

500 SHARES

Golden Zone Mine, Inc.

FULLY PAID	ONE MILLION SHARES NO PAR VALUE	NON-ASSESSABLE

This Certifies that R. L. Johnston is the owner of Five Hundred Shares of no par value each of the Capital Stock of

Golden Zone Mine, Inc.

...nsferable only on the books of the Corporation by the holder hereof in person or by Attorney upon surrender of this Certificate properly endorsed.

In Witness Whereof, the said Corporation has caused this Certificate to be signed by its duly authorized officers and to be sealed with the Seal of the Corporation at Anchorage, Alaska, this 2nd day of Nov A.D. 1939

Actg. SECRETARY

PRESIDENT

SHARES NO PAR EACH

Figure 31. Common stock certificate, Golden Zone Mine, Inc., 1939. "R. L. Johnston is the owner of Five Hundred Shares. . . ." Courtesy, R. L. Johnston, Anchorage.

dent, Alex McDonald, had come to Alaska in 1900 for the Nome gold rush and had worked for Alaska Steamship Company since 1906. In 1938 he was its Anchorage agent, a director, and its Alaskan manager from his building on Fourth Avenue, the building that now houses the 515 Club. "Tar" Tarwater had not stampeded to Nome, but he was in Seward by 1916. Before coming to Alaska, he had served the federal government as Commissioner of Education in the Philippines. Tar's great extracurricular civic claim to fame was founding the Anchorage Golf Club and developing the city's first golf course.[18]

A fourth man who was also deeply involved in GZM—A. A. Shonbeck—was not an officer, but he was a director and sometimes Dunkle's chief supporter. Shonbeck was Tarwater's and McDonald's contemporary, about ten years older than Dunkle. Born in Wisconsin in 1878, Shonbeck had stampeded to Nome in 1900 and had participated in the Iditarod rush in 1910 before moving to Anchorage in 1915. By the 1930s Shonbeck's interests were so well known that his advertisement in the *Anchorage Daily Times* was simply a bordered space that said "A. A. Shonbeck." Shonbeck was also a civic leader and political power in the territory. Many of Dunkle's Anchorage friends could be classified as members of three groups: the Presbyterian Church, the Elks Club, and the Republican Party. Shonbeck was an Elk, but he was a Democrat, who had served as Democratic National Committeeman for Alaska.[19]

The mining leader, Dunkle, and his three key men, McDonald, Tarwater, and Shonbeck, were a pretty unlikely bunch of mine promoters. Some of today's mine promoters in Vancouver or Toronto, Canada, wear heavy gold neck chains or sport pockets full of nuggets. There was not a neck chain of any variety among Dunkle's backers, but they did promote. They also did something that many stock promoters do not do. They violated the OPM (Other People's Money) rule by investing substantially with their own dollars in GZM. The promoters believed in the project. Dunkle invested $100,000 in GZM common stock, and McDonald had at least $20,000 invested. Mr. and Mrs. Shonbeck owned nearly $6,000 worth of common stock, and they may have owned more through loans protected as preferred shares.

A nearly final accounting in 1945 showed that GZM had sold more than 880,000 shares of common stock to almost five hundred people, mostly in Anchorage. The shareholders were diverse and included one of Anchorage's earliest pioneers, Oscar Anderson, who came in 1915, and an organization of "pioneers," Igloo No. 15 Lodge of the Pioneers of Alaska. Businessmen—D. F. Hewitt, Harry Hoyt, Charles J. Odermat, Milt Odom, and Bill Stolt—were substantial investors, as was Fritz Nyberg, the chief ranger of Mt. McKinley National Park. The corporation appeared to be nondenominational. It included two Anchorage businessmen and civic leaders who were Jews: Ike Baylis and Zac Loussac. There was no apparent discrimination against white women—Lillie Waterworth owned more shares than Frank Waterworth—or against black women—Anchorage madam and otherwise entrepreneur Zulu Swanson was a GZM investor. Doctors invested; some, including Dr. A. S. Walkowski and Dr. P. I. Heitmeyer, invested quite heavily. The dentists were represented by Lawlor Seeley. A young Anchorage banker, Elmer E. Rasmuson, who later became a multimillionaire, owned ten shares, which were initially worth four dollars.[20]

The mining project was adequately funded by stock sales from 1938 through early 1940. Sales in late 1938 and early 1939 were consistent, and for a depression-era project, the numbers are surprising. On 1 December 1938 three Alaskans invested $1,050; on the second, $720 was subscribed, including a contribution of $400 from Fairbanks journalist Kay Kennedy. On the third, the day's total was $550, and on the fifth, $815. Very few days showed less than $100 in sales.[21]

A continuous infusion of funds was needed because one part of the project was costing more than anticipated. Miners were driving the main haulage level five hundred feet vertically below the top of the ore body. The tunnel, which would allow the transport of ore from the top of the mountain into the mill building, was planned to be at least two thousand feet long. Dunkle had estimated the cost of supplies and equipment for the tunnel at about fifteen thousand dollars, but the ground conditions were terrible, and the costs soared for both labor and timber used in ground support. Dunkle, always in good graces with Harry Townsend of Anaconda, even after the company's pullout from the Golden Zone project, wrote Harry that timbers were jammed solid against each other at the tunnel's portal to support the ground in a huge fault zone.[22]

The ground improved on the other side of the fault, but conditions were still worse than anticipated, and costs were higher. At about nine hundred feet into the mountain, the miners got a temporary reprieve. In a 1938 report to shareholders, Dunkle predicted that the tunnel would intersect ore before it reached the main breccia zone. In June 1939 miners began to break into ore stringers in the face of the tunnel, and the assays began to improve. The favorable news coincided with the annual meeting of GZM in early July. It was the best news that shareholders had received in several months. Better news followed when Dunkle returned to the mine a few days later. The ore stringers had consolidated into a solid vein that filled the tunnel. The discovery was announced in the *Anchorage Daily Times* on 20 July 1939: "Extremely rich material at least 8 feet wide found in tunneling."[23] The article continued proudly, "The project has been financed almost entirely with local money, only a small part of the stock being sold in Seward. There is no capital from the States involved." Dunkle immediately shipped a bulk sample of the vein material to the Tacoma smelter for testing. Within a few weeks the smelter reported that the ore contained 1.28 ounces of gold per ton, 14.52 ounces of silver per ton, and 15 percent combined copper, lead, and zinc. The smelter noted that it would be a difficult ore for them to handle, but that it could be smelted at an ASARCO plant in Helena, Montana.[24]

After the vein discovery was announced to shareholders and then to the public, share sales exploded from both insiders and outsiders. During mid-August, sales were so strong that Tarwater told the Fairbanks Chamber of Commerce that all available common stock had been placed. More stock was found, however, and sales maintained a goodly pace throughout 1939. Although the funds were expended about as fast as they were received to pay for the heavy costs of construction and mine development, it was a tremendous show of support for Dunkle. As 1939 closed, production at the Golden Zone mine seemed imminent.[25]

Dunkle had begun work at the Golden Zone mine in 1936, when it was an exploration-stage project without the assurance of an ore body. His faith in the project was vindicated after the breccia ore body assumed a form and after the miners found rich outlying deposits, such as the vein discovered in driving the haulage level. Dunkle's confidence was not destroyed when Francis Cameron and Anaconda Company rejected the project. Cameron wanted more assurance of rich ore than Dunkle did. Although it had not been proven by drilling, Dunkle was confident from his geologic projections that the rich ore was there. He was also optimistic that he could mine and mill low-grade ore at a substantial profit, if necessary.

After the decision in May 1938 by Dunkle and his directors to develop the Golden Zone mine, the project had to move quickly toward attaining cash flow through production. The mine began production in January 1941 and shipped ore concentrates until January 1942. But, as built, the Golden Zone was not a profitable mine. It probably would have ceased operation even without the draconian influence of World War II. What happened to Dunkle's dream?

In retrospect, the answers are easy but convoluted. Fundamentally the project was undercapitalized, but the extent of undercapitalization has to be viewed against an economy moving rapidly from depression to wartime inflation. When the project began, men were eager to work for room and board and a small paycheck. By 1940 unskilled men could earn more than Dunkle's skilled miners and mechanics as the federal government began to construct defense bases and airfields throughout Alaska. Geologic conditions in the mine were more difficult than anticipated and hard to estimate in budgeting. The mill was inadequate, not through a lack of knowledge, but from a lack of capital.

As production began, two problems surfaced: the delivery of ore from mine to mill and the recovery of enough metal at the mill. A third critical area involved generating and delivering electrical power to the mine and mill. The solutions that Dunkle and his comanagers adopted to solve the problems in ore delivery and milling were expedient. They were the best that could be found with the equipment and funds that were available, but they were not adequate to allow profitable development of the mine. Dunkle's solution to power generation was arguably the best and most efficient solution. It was inventive, but it was also controversial.

Dunkle's key to efficient low-cost mining was going to be the haulage tunnel driven from the mill to a point underneath the breccia ore body.

In the early days of the mine, high-grade ore could be mined selectively from the upper workings, dropped to the tunnel level, and hauled to the mill by rail. The tunnel would be driven underneath a *glory hole* (see glossary), where low-grade ore would be pushed from the walls of the hole and would fall to the haulage level. It was a mining method that Dunkle had used when he was the short-term superintendent of the Beatson mine on Latouche Island. At the Golden Zone mine, he proposed to modify the method to some extent by using mechanical shovels and by eliminating the timbered ore chutes used at Latouche. In 1938 Dunkle told his shareholders, "This should avoid a great part of the heavy preliminary development of the Latouche system, should cost little more from the operating standpoint, and will give us the opportunity to do some rough sorting with the shovel. This feature I consider important on account of the large horses and dikes of hard diorite and porphyry that occur in the ore body."[26] The plan had to be abandoned, however. Because of the cost of driving the haulage tunnel through difficult ground, the tunnel was stopped about eight hundred feet short of the breccia ore body, not long after it had intersected the high-grade vein that had caused the rush on the GZM shares.

Something of the original plan was retained when miners drove a mill hole or transfer raise to the surface from the end of the tunnel. The ore could then be trucked from the upper mine workings and dropped down to the mill level, where it would still have some protection on its way to the mill. It worked, but the ore had to be transferred several times instead of once or twice; in miner's parlance, "the ore was worn out" in transit. A road had to be maintained during the winter between the 200 level and the transfer raise. It raised hob with Dunkle's projected glory-hole mining costs, which he had earlier explained to Townsend: "We shall push the ore into these raises with a ninety horse-power bulldozer. . . . As the longest push shall not exceed 200 feet, and the ore is easily broken . . . , we expect to put the ore into the raises for not to exceed 20 cents per ton." He figured it would cost another five cents for rail haulage in the tunnel.[27] Undoubtedly Dunkle's projected mining and haulage costs rose by a factor of several times.

Problems with gold recovery at the mill were related to cost, availability of equipment, and consequent departures from Leo Till's well-designed mill flow sheet of 1938 (fig. 32). In Till's original plan the ore would go through two stages of crushing before it arrived at the ball mill. Because of a lack of funds, the mill had only one crusher, and coarse material was fed directly into the ball mill. The purpose of a ball mill is to grind ore to a powder, a size necessary for further processing,

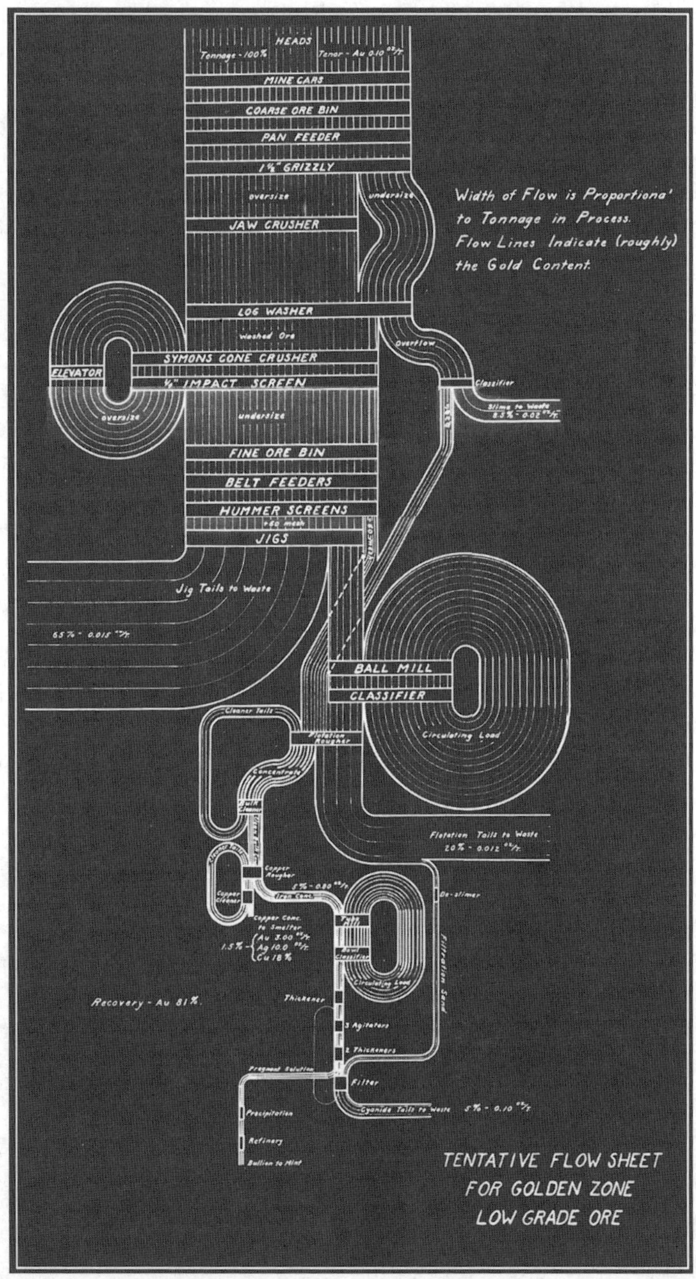

Figure 32. Flow sheet for planned Golden Zone mill, 1938. The mill as built lacked features of this design and consequently recovered less gold. From WED, "Golden Zone Mine: Report to Shareholders."

but without second-stage crushing, the mill could not keep up with the coarse feed. Dunkle proposed a partial solution that depended on the ore itself. Hard quartz-rich parts of the ore were generally low grade and gold poor; most of the rich ore was much softer, and hence, more easily ground. Till added a coarse screen to the outfall of the ball mill, where the smaller and hopefully richer fragments passed through the screen for further processing. The coarser, presumably poorer, hard quartz fragments went directly to the tailings pile. The screen addition allowed the mill to keep up with the crusher, but occasionally large fragments of gold-bearing ore were discharged into the tailings; all of the coarse tails contained measurable amounts of gold.[28]

An equally serious modification to the original mill design was the substitution of single-stage for two-stage flotation. Till originally proposed to concentrate all of the sulfides in one flotation step and then selectively float off the copper in a second step. The copper-rich concentrate, which contained a high percentage of gold and silver, would be shipped to Tacoma for smelting. The remaining low-grade, iron-rich rock would be leached at the mine to recover more gold. Instead the mine delivered a relatively low-grade bulk-flotation concentrate to Tacoma for smelting, and no gold was recovered at the mine. Consultant Alexander Smith later determined that the Golden Zone mill recovered only about 55 percent of the gold instead of the 81 percent anticipated from Till's original flow sheet.[29]

A key component of Dunkle's original plan was the on-site generation of hydroelectric power, an item with a moderate initial capital expense and a low operating cost. Hydroelectric development at the Golden Zone mine had actually begun in 1936, when Dunkle installed a small Pelton-wheel generating plant in lower Bryn Mawr Creek for seasonal power development. Later the flow down little Bryn Mawr Creek was augmented by bringing water from Long Creek, the next main drainage to the south, and a larger plant was installed. Bryn Mawr and Long Creeks, however, only supplied power during the relatively short open season, from about June until mid-October. During the nearly eight-month-long winter, power was supplied by diesel electric generators, a cost that Dunkle had hoped to avoid. The search for an all-season hydroelectric source began in the fall of 1938. Dunkle explained the developments to Harry Townsend: "We are putting our hydro-electric plants down on the West Fork. . . . As you know the stream is a glacier stream . . . ; the gravel is loose and saturated with water. I had a hunch that bedrock would not be very deep and that it would carry a great deal of water, so we dug a bedrock drain . . . , starting it at the head of a ditch on the opposite side of the river from the Riverside [prospect]." The ditch, then in construction, was more than two miles long and had a 1 percent grade, or 125 feet of drop in 11,000 feet of ditch. Dunkle's hunch proved right; he continued: "[We] struck a false bedrock of clay and gravel at a depth of 11 feet. . . . As expected there was a heavy flow of water on this bedrock, and the face of this cut, about 4 feet wide, produced about 1 second-foot of water when we quit there last fall."[30]

Shortly before he wrote his letter to Townsend, Dunkle had dug through the midwinter snow and had found about the same amount of water. Discounting possible lower flow rates in the center of the broad glacial valley, Dunkle expected to produce "close to 100 second-feet with a drain eight hundred feet long and that will give us 1000 horsepower at our head. The water will be warm in the winter and perfectly clear at all times, and the miles of gravel above us will make an ideal storage basin."[31]

Thanks largely to the shutdown of the Kennecott mine in the Wrangell Mountains, the Golden Zone mine had a fully equipped machine shop. Dunkle was at the Kennecott mine buying surplused machinery while the last ore was cascading through the mill.[32] At the same time, Dunkle also obtained the services of master mechanic Carl Engstrom, electrician Clarence O'Neal, and a millwright probably named Johnson: men who had decades of Kennecott experience. At the Golden Zone mine, Engstrom was capably supported by another excellent mechanic, Rex Anderson, and by blacksmith Al Dolan. These men could fabricate just about anything that was needed at a remote mine. Using their practical skills, Dunkle proceeded to the next part of the hydroelectric plant: designing a hydraulic impulse wheel to generate the power. He began his design theoretically by calculating, via algebra and calculus, the best angle for the water jet that would drive the wheel. He then designed the wheel itself, which he believed was a significant improvement over previous impulse wheels (fig. 33). As he told Townsend,

> I have invented a water turbine that is an entirely new design as far as I know. It is a small wheel of high speed under a low head. For instance a wheel to develop 500 horsepower under 125 feet of head would be 15 inches in diameter and turn at 1500 R.P.M. . . . I have made up a model that turns 1000 R.P.M. at four feet head and develops 1/4 H.P. I am going to make one that develops 5 H.P. at 30 feet head and 1750 R.P.M. and connect it to our flume (we have a 150 H.P. development on this side of the river at 30 feet of head) and try it out.[33]

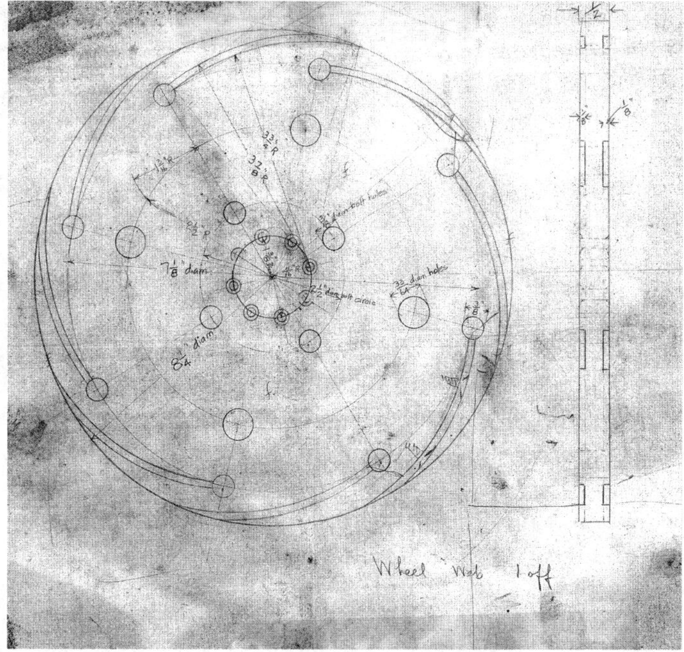

Figure 33. Dunkle-drawn construction diagram for a small-scale hydraulic impulse wheel, 1940. Author's collection.

Dunkle also anticipated a low operating cost for the system, as he further related:

> To go with the turbine I have developed a power absorption governor that is mounted directly on the water wheel shaft. . . . It is an air-fan device and works perfectly in the model. The only trouble with it is that in large powers it would stir up a great commotion, but that, again, would be no serious drawback out here in the wilds. . . . With this combination and the proper electric controls and proper lubrication storage I believe that we could leave the plant for a week or more at a time to operate itself, but what I am really aiming at is a plant that would require only one operator for 24 hours.[34]

Dunkle also designed modifications and options for the system to compensate for different flow rates. Perhaps the air-fan system proved to be too noisy even for a remote site because the final plant used a simple water-float regulator to adjust the flow to the wheel. The water-supply ditch was completed as a drain on the hardpan layer for more than a half mile across the West Fork; it was piped with porous wood culvert and then backfilled. It supplied the already completed ditch on the south side of the West Fork (fig. 34). The system was connected to a hydroelectric plant that was dedicated in August 1940.

The hydroelectric plant generated most of the mine's power during its brief period of operation. In 1967 Dunkle's dreams on power generation were still a subject of discussion and speculation along Alaska's railroad belt. Various second-guessers proposed that a diesel-electric or coal-fired power plant would have been a better option. Nevertheless, these solutions would have had their own characteristic problems and costs. Particularly in the low cash-flow year of 1941, "free" power must have seemed an asset at the mine.

The total impact of the problems at the Golden Zone mine was not felt until 1941, when the actual mining began. In the meantime, life was busy, regular, and often fun. The only holidays that were taken were Christmas and the Fourth of July, but Dunkle was flexible with his experienced crew. If a man needed more time off after a holiday, his job was still there when he returned to work. Everyone had prospecting privileges, to the extent that they did not conflict with some critical job. Dunkle's leadership abilities, which were first recognized by Kennecott at the Midas mine near Valdez and at the Beatson mine, remained clearly evident at the Golden Zone mine. The mine crew—both miners and bosses—followed Dunkle from the Lucky Shot to the Golden Zone, and the men were willing to draw half their pay in GZM stock. Admittedly the payroll scheme conserved cash, but the miners and Dunkle saw it as spreading ownership and making capitalists out of laborers. Everyone had a common incentive, as recognized by Fairbanks journalist Kay Kennedy when she visited the mine in July 1939: "Forty men, all of whom have a monetary interest in Dunkle's Golden Zone mine near Colorado, are feverishly working."[35]

Bachelors, by far the most abundant employees at the mine, lived in a well-constructed three-story bunkhouse. Married staff lived in little one-and-one-half-story houses finished in basic tarpaper black. The residents of three staff houses were electrician Clarence and Mrs. O'Neal, mill man Leo and Anna Marie Till and her two sons, and mechanic Rex and Mrs. Anderson. Earl, Billie, and young Bruce Dunkle lived in a smaller black house just below the staff houses. Cecil Wells, the nephew of Golden Zone discoverer Lon Wells, and his wife, Paula, who were favorites of Dunkle, lived in a log cabin just below the garage and the cookhouse.

Figure 34. Construction of the buried collection drain under the West Fork of the Chulitna River (top) *and the completed water-supply ditch* (bottom), *Golden Zone mine, ca. 1940. Courtesy, respectively, Dunkle family, Lu Liston photograph, GZ-95; and Anchorage Museum of History and Art, Lu Liston collection, 1018.3.*

Miners, shift bosses, and unmarried staff ate their meals at the big cookhouse. The Dunkles and married staff usually ate breakfast and lunch at home, but everyone ate suppers in the main dining room. The food was plentiful and good; Dunkle and his friend Henry Watkins had long agreed that the cook was the most important person in a mining camp. Alaskan mining camps had a tradition of friendliness and a rough but not mean camaraderie. The Golden Zone camp was no exception. There was always room for occasional guests from Anchorage or Fairbanks, such as Dr. and Mrs. Paul Heitmeyer or Lu Liston. Dr. Heitmeyer was the doctor for the Alaska Railroad; his wife Winifred was a special friend of Billie; Lu was a friend and commercial photographer, who took many pictures of the mine as it was being built.

Rare holidays were celebrated. At Christmastime, Billie Dunkle decorated the large dining room and promoted seasonally appropriate entertainments. She could not make old bachelor miners sing about good King Wenceslas, but she was successful in promoting a horse-racing game, in which everyone could bet on the outcome and easily recognize the participants. Jockeys and horses bore humorous monikers; jockey A. N. Ville rode the blacksmith's horse named Forge Ahead, and the electrician's jockey was Killer Watt.

After 1937 the mine was connected to the Alaska Railroad by a passable roadway and series of bridges, and Dunkle had a Stinson airplane, which he used for light freighting and for picking up new hires. He continued to have a few adventures in the air, some with less than serious consequences. One such incident was described by his young plumber, Luther ("Tex") Noey. The men had just taken off from the river bar when the engine quit. Dunkle said, "We're going to crash." Noey barely had time to say, "We sure are," before they did.[36]

Dunkle's main task, however, was financing an almost completed mine. Dunkle initially planned to obtain some of the money from the RFC. Based on preliminary discussions with RFC, the project could have received a twenty-thousand-dollar development loan. Dunkle, however, believed that the loan should be deferred and that developments should be financed by additional sales of stock, as he explained to his shareholders:

> In the original plan for development and equipment an application for a fairly large RFC mining loan was contemplated, preceded by a development loan of $20,000. The larger loan could not have been made without the expenditure of the development loan. The only available place to spend the money was in the upper workings in ore. . . . Such development work would serve no useful purpose. . . . It was, therefore, decided . . . not to apply for any loan at this time but to raise additional money by stock sales and drive the lower tunnel into the better ore . . . indicated by diamond drilling. This tunnel will serve not only to extract the ore below the upper workings but will constitute the haulage-way for all the [upper

> level] ore to be mined by surface methods. . . . With this ore in sight, and the construction and development so far advanced, an application can be made for a mining loan without the preliminary step of a development loan. The amount of money needed will be much less, and it is believed that an application made then will be accepted.[37]

There were some who disagreed with Dunkle's analysis. One was C. F. ("Chuck") Herbert. As a fledgling mining engineer, Herbert had conducted assays for Dunkle at the Lucky Shot mine. By the late 1930s Herbert was an examining engineer for the RFC in Alaska. He believed that Dunkle should apply for a development loan and that there was sufficient flexibility to allow for the money's efficient use. In this instance Dunkle's certainty in his own plans ultimately hurt the project. In a last settlement letter written in 1940 for his work at the Lucky Shot mine, Dunkle appealed to Baragwanath of Pardners Mines for the names of individuals who could supply twenty thousand to thirty thousand dollars, amounts which then could have been of crucial value to the project. In 1939, however, Dunkle and most other Americans still could not foresee the coming firestorm of a world war or even the end of the depression that had given Americans the option of RFC loans. Nevertheless, the oncoming war began to affect almost everyone. The effects of the war were more noticeable in Anchorage than at the Golden Zone mine. Billie's friend Winifred Heitmeyer informed Billie, "With bombers arriving and departing we get a reminder of what the same means in England. The air base work is being pushed with all possible speed—One wonders at times—why all this rush?"[38]

By the winter of 1940, surface construction ceased. The mill complex was as complete as possible based on the amount of capital available for the project. Although there were internal deficiencies, the resultant project was impressive. The mill-haulage portal area was nearly covered by a complex of shops and railways that were protected by snow sheds and were tied to the steep hill in case of avalanche. The mill itself was a sturdy three-story building constructed on concrete piers and footings (fig. 35). All of the buildings were made of wood, but as much as possible they were protected against fire by extinguishers, water-filled buckets, hoses, and steam-heated water lines from an insulated building that contained a huge cylindrical water-storage tank.

Underground development work continued throughout the year. The miners, led by shift bosses Crockett Metcalf and Tony Butorac, opened up development headings and prepared stopes for mining. Because of rapid employee turnover, more than one hundred men eventually worked

Figure 35. Golden Zone mill and shops at the mill haulage level, ca. 1940. The steam plant is in the valley to the right of the mill. The railway in the lower left is for mine waste. A covered railway extended from the haulage level into the mill. Courtesy, Dunkle family.

in the mine crew; a maximum of sixty worked during any one month. Tune-up of the Golden Zone's mill circuits began in October 1940, several months after the mine's projected opening, and the first concentrates were shipped in January 1941.[39]

Dunkle regarded the operation of the mine through 1941 as experimental, although at least 869 tons of concentrates were produced for shipment to Tacoma, probably from about eight thousand to ten thousand tons of ore.[40] After shipping several loads of copper-bearing concentrates that were fairly rich in gold, Till and Dunkle decided to ship bulk concentrates that contained less gold but more arsenic. Most of the concentrates shipped later in the year were nearly pure arsenopyrite; they contained about two to three ounces of gold per ton, ten ounces of silver per ton, and more than 25 percent arsenic.[41]

On 7 December 1941 Japanese bombers answered Winifred Heitmeyer's rhetorical question—"Why all this rush?"—when they struck an almost idle U.S. Pacific fleet at Pearl Harbor. The bombers indirectly killed an Alaskan mining project that was already infected with serious undercapitalization. The mine closed in January 1942.

11

Placer Gold, Coal, and Postwar Reality

Dunkle's other late-1930s venture started after the Golden Zone project, attained production sooner, and was more successful. It was a placer gold mine, which was inherently simpler than a lode mine. For its time, however, the placer mine at Caribou Creek was an advanced project using the best equipment available. Operationally the project meshed with that of Golden Zone. With the aid of his airplane, Dunkle could manage both projects from his base at Colorado Station on the Alaska Railroad.

Caribou Creek drains the north flank of the Kantishna Hills, the locus of the Kantishna mining district, which contained many intriguing lode mines of gold, silver, and antimony (fig. 36). When Dunkle began his exploration in 1937, Caribou Creek already had a long mining history. In the early years, miners developed rich, shallow, gold placer deposits that they could mine by hand. In the 1920s William Taylor, one of the miners who made the first ascent of the south peak of Mt. McKinley, promoted Caribou Creek and tried unsuccessfully to mine it as a large-scale hydraulic project. Shortly afterward, placer-mining consultant C. A. Bigelow examined the area for W. P. Hammon of California. While drilling the property with a churn drill, Bigelow found a large but low-grade resource that was not economic at its site, and he rejected it.[1]

Dunkle reevaluated the Caribou Creek project after the price of gold was raised. Besides noting the positive effect of the higher price, he believed that the grade of the prospect could have been underestimated because it was based on gold recovered from small-diameter holes that were churn-drilled in thawed ground. The economics of the project had certainly improved because, as of 1937, there was a rough road into the area from McKinley Station. Dunkle optioned the William Taylor property. His friend and business associate, Glenn Carrington, advised him that other properties along the creek were also available. Dunkle finalized

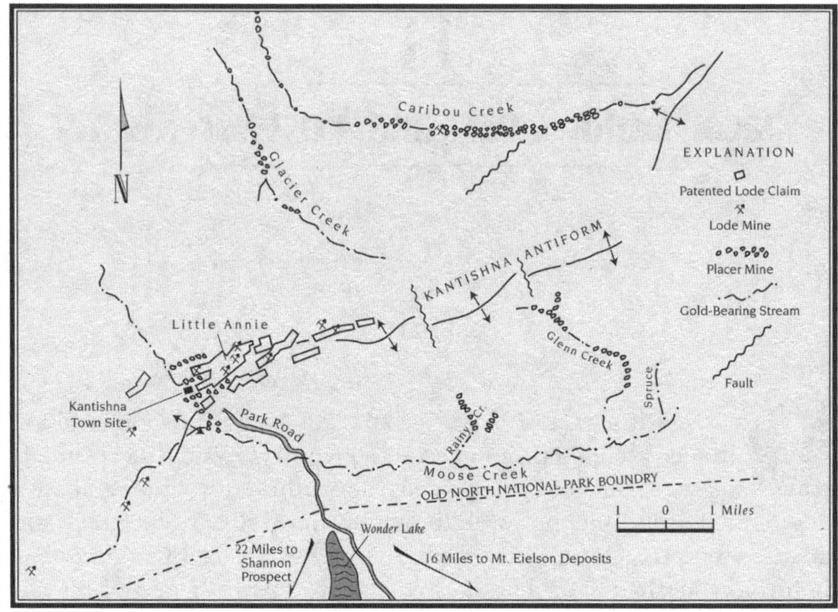

Figure 36. Kantishna mining district, showing the locations of Caribou Creek, Little Annie, and other mines. After Kantishna map in Hawley and Associates, "Mineral Appraisal."

other leases, including some with young Fairbanks miners Ernie Maurer and Meehling, who had a sublease from Taylor and who controlled other ground along the gold-bearing creek.[2]

By the summer of 1938, Dunkle was prospecting the creek and sharing his expenses with Carrington. In his reevaluation of Caribou Creek, Dunkle used the caisson system, in which a large-diameter lined hole is sunk through the deposit, yielding a much larger and more representative sample than through churn drilling (fig. 37). The caisson samples indicated more than twice the value obtained by Bigelow. Dunkle and Carrington decided to proceed and set up a business partnership that also included L. C. Thomson, owner of the Lucky Shot mine, and, for a half interest, Dunkle's Seattle banker, E. B. Kluckhohn.[3]

The balance of 1938 and the early part of 1939 were spent in planning the operation and ordering equipment. Most of the detailed work was handled by Dunkle or by Arthur F. Erickson, Carrington's partner. The men agreed from the start to use some kind of elevated sluice box fed by a dragline; they disagreed, however, on the complexity of the

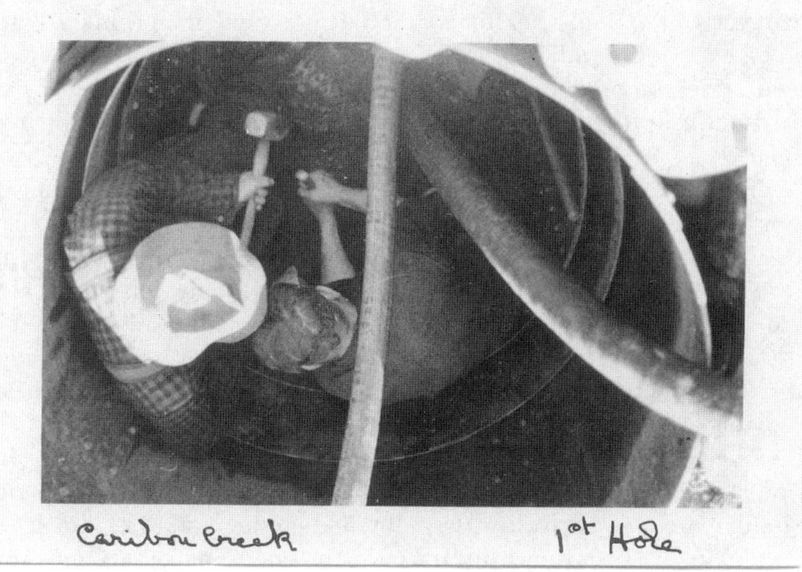

Figure 37. Caisson prospecting for placer gold in Caribou Creek, Kantishna district, 1938. The caisson system yielded larger and more representative samples than did previously used churn drilling. Courtesy, Dunkle family.

plant and on how much and what kind of equipment to use during the first year's operation. Erickson was more conservative than Dunkle. Based on his experience with operations that had failed because of fancy equipment, Erickson opined, "I would be inclined to take more or less a standard outfit up there that would fit the job and then branch out from there."[4]

The debate over equipment was resolved because Washington Iron Works of Seattle had just designed a mobile *washing plant* (see glossary), which was essentially a land-based dredge. Dunkle and Erickson, who were both cautious before, were enthused about the new plant. Erickson wrote, "[T]hey really have arrived at a good-looking washing plant at last. . . . When you consider the cost of the dragline for stacking tailings, also the expense and weight and freight of an elevated trestle, sluice box etc., this does not look too bad."[5]

The partners ordered the new plant, which was essentially a prototype at a special price of eighteen thousand dollars. Erickson worried whether it would arrive in Alaska by the summer of 1939, but it left Seattle on Alaska Steam with just enough time to make it. On 3 April 1939 Carrington wrote George Campbell, Dunkle's office manager at the Golden Zone mine: "It was very gratifying to get your report on the equipment going to Caribou. . . . [W]e surely feel that we have assembled the most modern outfit ever to be shipped North for placer mining. If you think your prayers will be answered, we hope you will do a good job of praying for a late breakup, so that we can get the equipment all on the property this spring. We received a wire to the effect that one load had reached Caribou and that they were doubling up on the equipment, and we notice that the weather is still pretty cold in Fairbanks."[6]

Mobilization into the Caribou Creek area was a major job in itself. The dragline, the washing plant, diesel fuel, and all the supplies that were too heavy for air transport were off-loaded at the Lignite siding on the Alaska Railroad. Morris P. Kirk and Son, a major pre–World War II Alaskan bush contractor, was given the contract to move two hundred tons of equipment and supplies eighty-five miles to Caribou Creek. Kirk was contract mining at Earl Pilgrim's Stampede antimony mine, about fifty-five miles from the railroad, and was familiar with the rest of the route to Caribou Creek. Breakup held off long enough for Kirk to make the trip using two of his own tractors and both of Dunkle's project tractors to haul the freight sleds.[7]

With the heavy iron equipment on the trail to Caribou Creek, some lighter shipments were of concern. Groceries purchased in Seattle for the project were delayed when Dunkle failed to confirm a mid-March

Figure 38. Placer mining on Caribou Creek, 1939. Below left, *Arthur Erickson;* right, *Glenn Carrington. A one-and-one-half-cubic-yard dragline fed the washing plant at the rate of about fifteen hundred cubic yards per day. Courtesy, Dunkle family.*

shipping deadline, but they did leave Seattle on the SS *Baranof* before the end of the month, or at least most of them did. T. S. Emerson, manager of Alaska Brokerage Company of Seattle, noted that a shipping clerk had overlooked the order for currants: "Knowing how essential currants are for the cook we are sending a few packages by parcel post. . . . [T]here will be no charges for the currants."[8]

Caribou Mines Company worked nearly as planned during the first season. Experienced practical engineer H. G. ("Jack") Turnbull was engaged to manage the operations, leaving Dunkle free of the direct project work. Turnbull was an excellent choice. He had helped to erect the large bucket-line platinum dredge at Goodnews Bay and had no problem assembling the smaller equipment at Caribou Creek. The key to any dragline operation is the skill of the operators. Carrington found two experienced men, S. S. Brashears and A. McKenzie, who were scheduled to leave Seattle on the next sailing. Carrington also hoped to get an experienced miner, Herman Medford, to operate the washing plant.[9] Arthur Erickson helped Jack Turnbull with project management and, in midseason, showed the rather complex dragline and washing plant operation to his associate, Glenn Carrington (fig. 38).

The first gold bar from Caribou Creek, which contained 106 ounces of fine (pure) gold, was deposited with the U.S. Assay Office in Seattle on 18 July 1939. A larger bar (199 ounces of fine gold) arrived in Seattle on 27 July. Altogether eight gold bars—more than 1,650 ounces of gold—were shipped from Caribou Creek in that first season. Including an estimate for the gold that was left in the sluice boxes, the season's gross was about $55,000, an amount that was sufficient to pay $16,000 in wages and $8,300 in royalties and to repay 20 percent of the capital cost incurred during the first season. The venturers took no compensation but left $12,300 in the account to open the 1940 season.[10] The proceeds were probably not as good as expected, but they were undoubtedly much better than those from many other first-year operations known to Erickson and Dunkle.

Proceeds from the operation were also sufficient to pay other employee expenses, including social security and workmen's compensation insurance. Partner L. C. Thomson had recommended purchasing the insurance to ensure that the company would have sufficient cash reserves to care for injured employees. Kluckhohn obtained quotes of $4.71 per $100 of compensation for the operating employees and $1.25 for the cook. Thomson also urged insurance coverage for Dunkle: "A liability policy should include you, giving the company similar protection to that which they had at Willow Creek."[11] Workers' insurance occasionally proved beneficial. When Herman Medford fell off the dragline, he was taken to Fairbanks for treatment. Compensation could not help another employee, however. Fogelman, of unknown first name, should have stayed in camp instead of going to town. He was shot by a woman during a holiday in Fairbanks. The shooting appears to have been fatal because the company canceled a $105 expense check made out to Fogelman.[12]

Federal legislation enacted during the Caribou Creek project in 1939 caused Dunkle to make an unusual move. Dunkle seemed to take perverse pride in never having joined a professional mining organization. He was certainly eligible for the American Institute of Mining and Metallurgical Engineers (AIME) and the Society of Economic Geologists (SEG). He occasionally attended their meetings, but he refused to join either group. Legislative events, however, caused Dunkle to become a charter member of the Alaska Miners Association. Revision of the federal wage-and-hour law of 1939 pulled Dunkle and other Alaskan miners closer together than they previously had thought possible. As passed by Congress, wage-and-hour legislation was based mainly on industrial employees in the contiguous United States, many of whom worked regular hours five or six days a week for large corporations. Alaska's territorial delegate to the U.S. Congress, Tony Dimond, and the employers of commercial fishermen and miners, who worked long hours at remote locations in generally seasonal jobs, thought the law should include exceptions to cover their schedules. Dimond urged the miners to join together and send someone to Washington, D.C. to plead their case.[13] In March 1939, just when Caribou Creek was beginning to mobilize, Dunkle and other miners assembled in Fairbanks to organize the Alaska Miners Association, an organization that still represents Alaska's miners. The newly formed organization sent Bert Faulkner to the nation's capital to obtain concessions that reflected Alaska's employment practices. Dunkle's partner, Glenn Carrington, hoped to get a complete exemption for placer mining, but he recognized that this would be unlikely. He wrote Dunkle, "Until we find out what Congress is going to do about this amendment exempting anyone making $200.00 or more per month from the 44 hour week, it is rather hard to say about these wages, but they should be figured on a basis of so much per hour with time and a half after 42 or 44 hours per week."[14] (Dunkle's underground miners at the Golden Zone site were in a different position. They had been working a maximum eight-hour day for many years.)

Within a year, the rapidly approaching war would become very noticeable in Alaska, but in the summer of 1939, the main concerns at Caribou Creek were the amount of gold in the cleanups and the returns from the U.S. Mint. Production nearly doubled in 1940, when 2,935 ounces of fine gold were recovered. It decreased slightly to 2,373 ounces in 1941. The year 1942 was the best year; Caribou Mines Company produced 3,203 ounces of fine gold from more than 4,400 ounces of the silver-rich Kantishna placer gold.

The grade of mined material was substantially better in 1941 and 1942 than in earlier years. The amount of gold lying on bedrock was essentially the same, but in 1941 the operation had moved upstream to shallower ground owned by Bill Julian, so fewer yards had to be processed to recover the same amount of gold. Production was impressive during the 1940–1942 seasons. Caribou Mines Company cleaned about one million feet of bedrock per season and processed about fifteen hundred cubic yards of gold-bearing gravel per day.[15] Operations ran on a two-shift basis: five men on each ten-hour shift. The eleventh member and cook for the crew was Mrs. Schoentrip, the wife of miner Virgil Schoentrip of the production crew, who was a longtime Dunkle hand.

Caribou Mines Company did not operate between 1942 and 1946 because of the war. The mine reopened after the war, but Dunkle sold his interest to Carrington and did not participate. The operation at

Caribou Creek in 1946 was almost as good as it had been in the prewar era, but because of increased costs and a fixed gold price, it lost almost ten thousand dollars. Operations in 1947 and 1948 also lost money. Carrington finally regretted that he had reopened the mine after the war. Dunkle's decision to pull out proved prudent, but it apparently reflected a lack of funds more than superior judgment. Dunkle confessed to Carrington that the Dunkle family's expenses exceeded its income and that his decision to sell was a matter of money. Discounting a note to Carrington, Dunkle netted about ten thousand dollars on the sale, an amount that would have balanced the company's books in 1946 but was perhaps more valuable to the Dunkles.[16]

Wartime closure of gold mines such as the Caribou Creek operation was the rule rather than the exception. Most Alaskan gold mines ceased operations in the fall of 1942 because of War Production Board Order L-208, which banned mining operations that were nonessential to the war effort. Gold, a metal used mainly for money or jewelry, was deemed less essential than industrial metals, such as copper, nickel, and tungsten. A few gold mines, including Independence and the great Alaska-Juneau (A-J), remained in operation because they could produce minerals that were deemed essential or strategic as by-products of their gold operations. At the Independence mine in the Willow Creek district, the by-product was tungsten (from scheelite, or calcium tungstate); at the A-J mine, the by-product was lead (from galena, or lead sulfide). The A-J mine's lead production was considerable because of the scale of the operation: more than twelve thousand tons of ore per day. Operations were still difficult, however, for all mines. Materials were in short supply, as were miners, geologists, and engineers. Almost all of the mining professionals who were over the age limit for direct military service were involved in the war effort. Harold E. Talbott, formerly of Pardners Mines, was in Washington, D.C. managing aircraft production; his former associate, Jack Baragwanath, spent the war years in Cuba mining nickel.[17] Dunkle colleagues L. A. Levensaler, DeWitt Smith, Dave Irwin, Phil Wilson, and Ira B. Joralemon were either in Washington, D.C. guiding the logistics of converting ore into metal or in the field looking for strategic minerals. In addition to the metals, geologists sought quartz crystals and mica leaves that were clear and optically perfect. (The quartz crystals were cut into wafers to control radio frequencies; the mica leaves were used in bombsights.)[18]

Dunkle's mineral contribution to the war effort—coal—was less valuable than optical quartz but as essential as that strategic mineral. The big new military bases built at Fairbanks and Anchorage generated heat and power with coal, and the territory's existing mines were stressed to meet the demand. Dunkle's coal mine was only a few miles from the Golden Zone mine in the region now called the Dunkle Hills. Coal seams cropped out at the head of Costello Creek about five miles northeast of the Golden Zone mine. The seams had been mined on a small scale for decades, first by the Wells brothers, who used the coal for camp fuel and for the forges where they sharpened their tools. In the 1930s Henry Stevens obtained a lease (No. 09196) on the field and supplied fuel for the steam-boiler plant at the Golden Zone mine. By 1941, before U.S. entry into the war, the defense construction industry in Alaska had escalated so much that existing coal mines could not supply the demand. In September 1941, while the Golden Zone mine was still in production, Stevens, who was too old to start a new mine, transferred his lease to Dunkle, who made a new opening into the coal for defense-related production.[19]

Dunkle's coal production began soon after the closure of the Golden Zone mine in the winter of 1942. Three employee staff houses from the Golden Zone mine were skidded across the West Fork of the Chulitna River to the coal mine. The coal operation also used the shops, bunkhouse, and kitchen at the Golden Zone mine. Dunkle had a few employees, but because the coal was to be used in military bases, the army supplied miners. A seventy-five-man engineering company, commanded by a lieutenant and two sergeants (one "mechanical" and one "electrical"), was assigned to mine coal. It was not a particularly satisfactory venture. Dunkle later explained to Henry Watkins that he now understood the origin of the phrase "soldiering on the job." Soldiers who were willing to fight against their Japanese or German counterparts for twenty-one dollars per month were not happy with coal mining, and they avoided it whenever they could. It seemed almost like slavery to the men, and Dunkle was sympathetic to their plight.[20]

The Dunkle mine closed in March 1943 after mining about five thousand tons of coal. Although not highly productive for coal, the project had indirect benefits for Dunkle and GZM because the company had rented the rolling stock of the mine, including trucks, tractors, and a one-half-cubic-yard P & H shovel. Buildings and roads were maintained, always an important consideration in a heavy-weather environment. Dunkle also became a good coal miner, a skill that he applied during the last few years of the 1940s to pay the Golden Zone debt.

From 1942 until the end of the war, Billie and Bruce Dunkle remained in Anchorage except for one short period after the Japanese bombed Kiska in the Aleutian Islands, and Alaskans feared that Anchorage was a possible target. At that time they moved back to the safety of the remote Golden Zone site and stayed there until the threat of bombing had subsided. Billie made her own contribution to the war effort with stints at the USO Club in Anchorage. During one restless period, Billie considered leaving Anchorage for New York, where she could use her international background more directly in the war effort. Dunkle's more direct war effort supplied coal to the bases at Anchorage and Fairbanks, which never ran out of fuel but at times had less than a day's stockpile in front of their hungry furnaces.[21]

In 1942 and 1943, the thoughts of every American were on the war. In those years even victory was not assured. By 1944, however, it was clear that the Allies would be the victor, and Dunkle and his mining counterparts began to consider the postwar era. Dunkle began to plan a new coal mine in Costello Creek using larger equipment and experienced coal miners who had been released from the armed forces. He could not, however, forget the Golden Zone mine. His memories were shared by the staff of St. Eugene Mining Company, a subsidiary of Thayer Lindsley's Ventures, Ltd. St. Eugene decided to take a hard look at Dunkle's Golden Zone mine.

St. Eugene's man in the north was Alexander Smith, the well-known Canadian "quiet geologist." Smith wrote to Dunkle in early September 1944: "I have been advised by Mr. Ridgeway R. Wilson of our Vancouver office to contact you at Colorado, Alaska, and arrange to examine your Golden Zone property. . . . Am at present staying with O. M. Grant at his gold quartz property near Fairbanks. We expect to finish our work in 10 days & I could proceed to your property."[22]

Dunkle replied favorably, and Smith examined the mine, including the status of the property, the results of drilling and sampling by Dunkle and Anaconda, and the accounts of metallurgical testing dating back to the work commissioned by H. C. Carlisle in 1935. Smith quickly determined that the mill, as constructed, recovered only slightly more than half of the gold in the ore. However, he also accepted Dunkle's estimate of the ore reserves in the mine and Dunkle's suggestions for improving the mill to minimize its losses. Smith tended to favor Dunkle's interpretation of the value of drill core versus drill sludge: a dispute that was a main reason for Anaconda's decision to reject the property earlier. He was impressed with Dunkle's estimate of the miners' productivity in the wide shrinkage stopes. Smith also saw the potential for ore deep within the breccia ore body and agreed that the mill haulage level should be driven under the ore body as soon as possible.[23]

The property had substantial surface improvements, and thanks to Dunkle's timely purchases made at the recently closed Kennecott mines, the Golden Zone property had well-equipped shops. The mill building was sturdy with space to add more equipment and mill circuits. Furthermore, because of the nearby wartime coal-mining effort, the mine buildings had been occupied and maintained almost continuously since the Golden Zone mine shut down in early 1942. A few sections of the haulage tunnel had caved in, but these needed only minor repair. Smith said that ten thousand dollars would pay for the repairs to the haulage tunnel and for the purchase of some mining equipment.[24] The Golden Zone property looked very good to St. Eugene. It was a nearly developed mine ready to go back to work at war's end, which was approaching fast.

Smith recommended that St. Eugene acquire a controlling interest in the property. His immediate supervisor, Ridgeway R. Wilson, concurred, and Smith and Dunkle worked out an option agreement. Under the terms of the agreement, St. Eugene would put the mine into production at a minimum rate of three hundred tons per day and would modify the mill, essentially to the specifications shown on Leo Till's original flow sheet of 1938. They also mutually agreed to extend the haulage tunnel, which would eliminate at least two stages of ore handling and probe the deeper part of the ore body.

The basic structure of the agreement was a simple one that was based on net profits of the mine. When St. Eugene's initial capital of approximately three hundred thousand dollars was repaid, it would own 51 percent of the mine and its profits. Until St. Eugene's capital was repaid, GZM would receive 20 percent of the net profits, which would initially be used to repay its own debt and creditors. After St. Eugene had recovered its capital and GZM had repaid its debt, GZM would receive 49 percent of the mine's net profits. It was a straightforward agreement that appeared to protect everyone's interest, including GZM's creditors. Moreover, it was not a hurriedly conceived agreement because Dunkle had outlined its basic framework to his directors and interested parties in June 1944.

Seemingly the agreement needed only minor modifications, legal fine-tuning, and ratification by the corporate structures of GZM and St. Eugene. What followed, however, was a very complex series of negotiations by several parties, who were friendly but lacked the same interests. Part of the problem was due to the parties' physical locations. Dunkle was in Anchorage or at the mine. Ridgeway R. Wilson of St. Eugene was in Vancouver, Canada, and his man, Alexander Smith, was often on the move. The interested parties, including Glenn Carrington, banker E. A. Rasmuson, and Dunkle's associate, A. A. Shonbeck, were as likely to be in Seattle as in Anchorage. In terms of motivation, Dunkle, Wilson, and Smith were most interested in an operating mine or in identifying fatal flaws as early as possible in the development process. Carrington wanted the mine, but he also wanted to repay the debt and sales of new mining equipment. Rasmuson wanted more protection of his debt than he thought was given in the draft agreement. Shonbeck, who may have invested more dollars in the preferred shares of GZM than had Carrington and the Bank of Alaska combined, was a patient party, who tended to think optimistically (like a miner). At first time seemed of the essence. Carrington wired Dunkle that he was giving Wilson tentative terms for approval because Wilson had a directors' meeting in eastern Canada on 8 November. Carrington also asked Dunkle about a schedule for reopening the mine and about his availability to supervise the work. Dunkle replied by wire that he was available and that work could start soon. He also asked Carrington to wait for his [Dunkle's] airmail letters of 26 and 27 October before the terms were sent to Wilson. These letters are missing from the files, but a Dunkle letter of 1 November essentially restated the terms that he had worked out with Smith, namely the rate of production, the interests to be earned, and the order of repayment of capital. These terms were nearly identical to the ones that Dunkle had sent to his own directors the previous June.[25]

In the same letter, Dunkle noted that St. Eugene might have to advance some funds quickly to cover the balance of a first deed of trust owed to Lon Wells, the surviving discoverer of the Golden Zone deposit. Wells had just been diagnosed with cancer. Dunkle also cautioned his correspondents that because of the shortage of equipment and men, who were still tied up in the war, start-up of the mine would be slow. He preferred even a slow start with a few miners to no start because the project could easily accelerate as more men returned home. That Carrington could step somewhat out of line is suggested by a letter of 2 November: "[I]t was my [Carrington's] understanding that the deal would be made on the basis of $300,000 and it looks to me like it would be to our advantage to have them spend an additional $100,000 or $200,000 for equipment if the ore reserves justify a larger amount. While it would delay immediate revenue to the Golden Zone stockholders, it would surely increase their revenue a great deal if the property justified a 500 ton mill instead of a 300."[26] Carrington sold equipment, and he had less basis than Dunkle in projecting an initial mining rate. Dunkle also wrote to all the parties, citing additional material from his 19 June letter. Thanks to George M. Campbell's systematic accounting efforts throughout the Golden Zone project, Dunkle was able to summarize investments and debts incurred from the beginning of the project in May 1938. GZM's financial problems were serious but not crushing, and a great deal had been accomplished with the invested capital, which had been slightly more than $600,000.[27]

The items with potential near-term effects on the title of the property, which were a concern of St. Eugene, were the notes payable and the debt owed to Lon Wells, which was not listed in Campbell's summary. Notes payable of about $96,000 were owed to two parties: about $35,000 to the Bank of Alaska and slightly more than $60,000 to Carrington and Company (formerly Carrington and Jones). Both were chattel mortgages secured by equipment from the mine or mill.[28]

Dunkle's advance warning on Wells' condition had been justified. Wells, in fact, was being operated on during the same day that Dunkle wrote the letter about GZM's financial condition. The Wells family needed at least $3,000 for the operation. Dunkle had hoped to obtain the funds by a draw from the Bank of Alaska secured by $24,000 that was due in equipment rental from the coal project, but he was turned down. He borrowed $2,000, and his assistant at the Golden Zone mine, George M. Campbell, put up the rest.[29]

The significant problems with the St. Eugene agreement were apparently resolved by early December 1944, when a sales agreement was announced by both the *Anchorage Daily Times* and the *Fairbanks Daily News-Miner.* In early January 1945 the Fairbanks newspaper told its readers about Ventures, Ltd., the parent company of St. Eugene, and its founder, Thayer Lindsley. In addition to extensive Canadian holdings, Ventures, Ltd. had other properties in the United States as well as Bolivia, Peru, Nicaragua, and "other parts of the globe."[30]

Regardless of the strength of St. Eugene's parent company and the apparent fairness of the agreement, not all of the problems had been solved to everyone's satisfaction. Writing late in December 1944, E. A. Rasmuson set out a priority order of debt. His sequence was legal and contractual: first, the secured creditors, which were the Bank of Alaska and Carrington; second, the unsecured creditors; and third, the preferred shareholders. Rasmuson specified that the preferred shareholders could not touch secured assets until after 1 March 1946, when payment was due on the shares. The letter held a barely veiled club as to the order of priority: "[I]n fact I shall object to signing any agreement which does not maintain the same preference between the three classes of creditors as there now is and is above set out." The letter appeared to be unnecessary because Dunkle, Carrington, and Shonbeck did not disagree with Rasmuson's analysis. Within days Carrington responded to Rasmuson, noting his discussion with Shonbeck and expressing complete agreement with his position. Occasionally correspondence was broken with other news: Dunkle was in a train wreck outside of Fairbanks just before Christmas of 1944, but as usual he escaped injury.[31]

Two other divisive issues, which were technically more difficult than the creditor issue, soon emerged. One of these was the distribution of the depletion allowance; the other was the fundamental nature of the mining agreement between St. Eugene and GZM. Because he knew how critical both were to the project, Dunkle responded quickly. In his mind both problems were related to the nature of the ore body, which he knew better than any of his associates. He attacked the depletion allowance first. The allowance, a fixed percentage of project income, is exempt from income tax. A mine ultimately exhausts its own resource, a fact acknowledged by the depletion allowance. The problem was in a fair distribution of the allowance to GZM and St. Eugene. Dunkle explained the geological implications to his Seattle tax attorney, John Garvin:

> The Golden Zone mine is a low-grade one consisting of a core of better-than-average-grade surrounded by . . . and grading into material which is marginal both in grade and position and which will be ore only if the costs of production are brought low enough. To obtain the maximum amount of total profit from the mine as much as possible of this marginal ore will be treated. Therefore, the margin of operating profit per ton of ore will never be very great and it follows that the percentage depletion allowance will always form a large proportion of the operating profit. It is essential, therefore, that each party shall receive its proper share of the depletion allowance.[32]

St. Eugene's first reaction was to propose changing the net-profits-type agreement, which had been negotiated by Dunkle and Smith, to a gross-income-type structure, whereby St. Eugene would pay a fixed percentage of its ore sales to GZM. (The agreement type, although effectively a gross, is called a *net smelter royalty,* or NSR.) Dunkle opposed the change, using exactly the same arguments he had used on the application of the depletion allowance. Dunkle proposed that, because of the nature of the ore body, the total production should be maximized by mining to the lowest grade possible (the *cutoff grade;* see glossary). Royalty structures are definitely an important part of the cutoff grade because the NSR is taken off the top from the direct mine revenues, whereas the net profits are taken at the end. For many years, the NSR was the bellwether of the industry. Many, perhaps most, royalty contracts on large low-grade mines are now based on net profits, whereby total cash flow can be maximized by producing as much metal as possible. At the Golden

Zone mine, Dunkle advocated the large-scale, low-grade approach, and an ultimate 49 percent share of the profits, which was specified in the initial agreement, would have been a very fair return.

Tax and royalty issues were at least temporarily resolved or forgotten in early 1945, and St. Eugene was still being patient. Smith wrote to Dunkle: "It is to be expected that it would take considerable time to get the Golden Zone stockholders and creditors lined up."[33] At about the same time Carrington constructively suggested that St. Eugene's attorneys draft the final agreement. He also noted that both he and Shonbeck had concerns over the length of time involved in the negotiations between the companies:

> Both Art and I are anxious to get started with them [Ventures, Ltd.] as I know that they are having other deals put up to them from different parts of the country; and I am afraid they will get loaded up to a point where this may not sound as attractive as it did to begin with. We both believe that the important thing now is to get a release broad enough so that you can handle it to meet the situation when you come down. It would be a lot easier if the Rasmusons were here when Shonbeck could be with them, as I am sure that he can handle the situation with them better than anyone else.[34]

Finally on 1 March Dunkle transmitted to Shonbeck and Carrington a standby agreement, but he also expressed a fear concerning the Bank of Alaska debt: "I believe the greatest possibility of a hitch in the proceedings lies in Rasmuson's apparent determination to demand that St. Eugene build up some kind of a sinking fund to protect the chattels from damage through use." Dunkle pointed out that some of the debt to the Bank of Alaska had in fact been paid. He also noted that, far from being damaged, many of the chattel items were effectively inoperable and needed repairs that would be made during an active project.[35]

In early March St. Eugene brought up the depletion allowance again and proposed that it might be better to operate the mine solely under St. Eugene and pay GZM a royalty or to form a third operating company with shares distributed pro rata to St. Eugene and GZM shareholders. Dunkle again worked through the tax issue and may have satisfied St. Eugene's Wilson because, in a letter to Carrington a few days later, Dunkle still argued for the net-profit approach. The structure of the agreement, however, had been breached, and apparently both Carrington and young Elmer E. Rasmuson jumped into the gap with their own proposals. Dunkle brought up the matter with Shonbeck, who as usual was staying with the Carringtons. In a restrained letter, Dunkle closed with, "I am going to insist from now on that in any matter that lies between Golden Zone and St. Eugene that I be consulted first."[36]

On 22 March 1945 Dunkle tried to bring contractual matters to a conclusion when he wrote a letter to his creditors and shareholders. Dunkle accepted E. A. Rasmuson's concerns. He emphasized to all concerned parties that the option with St. Eugene could not proceed until creditors and stockholders had approved a standby agreement that was attached to the letter.[37]

A more important issue, the nature of the postwar economy, was raised by Wilson in June 1945:

> I am not in favor of starting work at the Golden Zone Mine until labor and supply conditions have greatly improved. The limited program at the Alaska Empire Mine [Admiralty Island] has proven to be very unsatisfactory from a labor point of view. Wages $1.25 to $1.40 per hour are, as you know, away out of line for the economic operation of relatively low grade gold mining projects, such as Golden Zone.
>
> I am sure that you will agree that it will be better to wait for improved conditions so that we can have a real chance to make a commercial success of your Golden Zone project.[38]

In his reply, Dunkle partially agreed, but he thought that more men would be available later in the year. He pointed out that wages should decrease in the long term because government-supported wartime construction was waning. Dunkle also indicated that he would find men who would be willing to do the tunnel work on a contract basis rather than a daily pay rate. He told Wilson that he had received signed standby agreements covering more than 90 percent of the debt and representing more than 75 percent of the shareholders. Correspondence from August shows that the deal was still very much alive but also that relations between GZM and the Bank of Alaska had worsened. Dunkle proposed that Wilson contact E. A. Rasmuson directly to outline the framework of a longer-term deal.[39]

A serious personal and economic loss to Dunkle and GZM was the tragic death of A. A. Shonbeck in June 1945. Shonbeck and his close friend and associate John Beaton, the discoverer of the rich placer-gold deposits at Flat, drowned a few miles from Shonbeck's Ganes Creek placer mine when their truck went off a bridge abutment. Shonbeck apparently had a heart attack and fell against the door lock mechanism of the truck, trapping the men.[40] Shonbeck and Dunkle had worked very closely together since the early 1930s and shared common interests in mining and aviation. Shonbeck had also been a stabilizing influence throughout the St. Eugene negotiations.

A satisfactory solution to the GZM deal with St. Eugene proved elusive, although a good meeting between Dunkle and Elmer E. Rasmuson, the son of Edward A. and incoming president of the Bank of Alaska (soon to be the National Bank of Alaska), went a long way toward resolving the differences between the bank and the mine. Rasmuson convinced Dunkle that the bank also desired a solution based on production, with little concern about the old debt.[41] By late 1946, however, the combination of World War II inflation and the fixed gold price had severely damaged the viability of the project in the eyes of St. Eugene–Ventures, Ltd., and they were unwilling to proceed.

With the loss of St. Eugene, Dunkle pulled out all the stops in trying to find a new partner for GZM. He continued to promote the Golden Zone mine to Kennecott through E. T. Stannard and was in touch with Anaconda through Harry Townsend. He also made some new connections. Dunkle's reputation enabled him to contact companies at management levels. One company he approached was Callahan Zinc-Lead Company. It was a logical contact because of Callahan's involvement in the Livengood placer gold property north of Fairbanks. Dunkle also approached the owners of the Alaska-Juneau mine. He thought that the Golden Zone mine, like the A-J, could be mined inexpensively and pointed out that Golden Zone ore was more than twice as rich as the ore from the dormant A-J mine, which did not reopen after the war. He pursued several other possibilities through companies and promoters.[42]

In 1947 Dunkle wrote a final report of record on the Golden Zone mine. In it he argued, "Although the ore is relatively low in grade certain favorable factors, inherent in the type of ore, the size and structure of the ore body, and an ample supply of cheap power assure that a large tonnage can be processed with relatively few men and supplies, thus assuring a very low over-all cost, even under present conditions or any that are likely to occur."[43]

Dunkle started his mining career in 1908, but he kept abreast of developments in mining and milling that would, in his view, make the Golden Zone mine economical after World War II. He was swimming against the tide. Although he could calculate a paper profit, there was very little interest in the project with gold priced at thirty-five postwar dollars per ounce. Gold was dead.

GZM remained in existence into the early 1950s. In 1948 Dunkle and his new board, composed of Cooper, Brown (first names unknown), and banker-attorney Warren N. Cuddy, worked to discharge the last obligations of the company. The Bank of Alaska and Carrington and Company both agreed to settle with GZM at 50 percent of their debt. Much of the debt to the bank and to Carrington was settled with the sale of equipment and mining consumables at the Golden Zone mine. Some of the equipment went to Emil Usibelli's coal mine near Healy, Alaska, which was then expanding for the Korean War.[44]

Although he was committed to the area near Colorado Station, Dunkle found time for a few projects elsewhere. One project that he liked because it was in the Willow Creek district, home of the Lucky Shot mine, was the Snow Bird prospect. Dunkle also tried to promote the consolidation of the entire Willow Creek district to Kennecott Copper Corporation and Newmont Mining Company. At Newmont Dunkle's old Yale friend DeWitt Smith pointed out to him that most of Newmont's own gold mines were closed because they could not operate profitably with the fixed price of the metal.[45]

The late 1940s were tough years economically for the Dunkles. Earl had only occasional consulting jobs and the revenues from a small-scale coal-mining operation near Colorado Station. Billie's efforts toward economic rehabilitation of the family were just beginning. The nadir of Earl's economic life probably occurred in 1950. He still owed a small debt to the Bank of Alaska. One of the last items of collateral was a stock of heavy iron balls, which had been used as feed for the ball mill at the Golden Zone site. The Bank of Alaska sold the ball inventory to the Independence mine in the Willow Creek district, which had just reopened after the war. It was, however, up to Dunkle to retrieve the balls. Dunkle went to the Golden Zone site by himself to retrieve every useable ball. He wrote to Elmer Rasmuson of the Bank of Alaska:

> Enclosed herewith is check for $468.70, which is the full amount paid, net, for the balls I shipped to the Independence. This is for 3.725 tons. . . . Handling all the balls was a tough job. I had to roll them individually down over a high bank and then had to build a trough and roll them down for five hundred feet further along the creek to the highest point to which I was able to bring the truck. . . . Anyway, I figure that I handled every ball in the shipment at least eight times by hand, one at a time.[46]

Earl Dunkle, however, was prepared to start over if necessary, and the Korean War gave him an opportunity. Dunkle was hired by the Army Corps of Engineers as an engineering draftsman, civil service grade GS-5, at a salary of thirty-one hundred dollars per year. He soon proved his worth for more complex assignments. The newly expanded army and air force bases at Anchorage had severe wintertime freezing problems. In the winter preceding Dunkle's arrival, the military had spent one hundred thousand dollars thawing and excavating frozen water lines. It was not too difficult a problem for a northern engineer to solve. Dunkle proposed taking waste heat from the cooling tower at the Ft. Richardson coal power plant and putting it into the water intake and supply lines, which solved the problem. Probably as a partial reward Dunkle was upgraded to mechanical engineer, GS-9, with an annual salary of forty-six hundred dollars in April 1951.[47]

In the meantime, Billie Dunkle, a tough and pragmatic Scotswoman, had been making considerable progress in reestablishing the family economy. Her first efforts—selling popcorn to soldiers in Anchorage and sketches in Seattle—were of limited scope and success but perhaps necessary as a business apprenticeship.[48] Those efforts were followed by more success in real estate sales, by a rather controversial venture selling military surplus with Z. E. ("Slim") Eagleston, and by a managerial job in the exchange at the Elmendorf Air Force Base.

Billie's short-term association with Eagleston raised a few eyebrows. Eagleston, sometimes called "Cast Iron Slim," owned night clubs, a military surplus store, and two major exmilitary scrap yards near the railroad tracks along Ship Creek in Anchorage. To his customers, Eagleston was a good guy, but he was also big and tough, with a temper to match his size. Shortly after Billie began working at the surplus sales office, Eagleston assaulted a city building inspector, was tried, and was sentenced to prison. Billie ran the operation while Eagleston served his sentence. Although part of Anchorage had looked askance at the seemingly strange partnership between aristocratic Billie and scrap dealer and gambler Eagleston, the arrangement was accepted by the Dunkle's oldest Alaskan friends, who were pragmatists and who may have remembered their own early rough days. Some younger friends, such as Maxine Reed (Mrs. Frank Reed Jr.), were proud of Billie for jumping into the financial fray on behalf of her family.[49]

Billie parlayed her next job with the 942-2 Branch Exchange, Elmendorf Air Force Base, into a move to Washington, D.C. The move provided new directions for her, her husband, and their son, Bruce. Billie, along with Bruce, left Alaska in October 1951, after helping to transform the Anchorage Air Force Post Exchange into an efficient business. She arrived in Washington, D.C. in mid-1952, after briefly working at post exchanges in Texas and Florida while in transit.[50]

Although she was jobless on her arrival in Washington, Billie had ideas and friends, Alaskan and otherwise, who had other friends and contacts. Some were also friends of her husband. One man, Harold E. Talbott, former head of Pardners Mines, was secretary of the Air Force; and Jack Baragwanath, also formerly of Pardners Mines, had good contacts almost everywhere through his friendship with Bill Paley of CBS.[51] Billie used the contacts to the best of her ability. A military intelligence staff job arranged by Baragwanath did not materialize, but Billie soon found a political niche and a friend who would stand her in good stead the rest of her life: Washington newsman J. Bernard ("Bernie") McDonnell. Billie joined McDonnell in McDee Services, a company that engaged in a variety of work centered on lobbying. By mid-1952 both McDonnell and Billie were caught up in the presidential campaign with staff positions in the Eisenhower organization.

In late 1952 Earl Dunkle temporarily left Alaska for Washington, D.C., and in the spring of 1953 it was Earl's turn to seek a new direction. He threw his hat sequentially into two rings: as governor of the Territory of Alaska and, when that position was filled, as president of the University of Alaska. Dunkle's candidacy for both positions had merit and support, but his support was atypical. In the competition for governor, then an appointed position in the U.S. Department of the Interior, Earl had support in the Eisenhower administration from Secretary Harold E. Talbott, but his principal references were Alaskans: Bob Bartlett, Tony Dimond, and Andrew Nerland. Dunkle knew that his references were rather unusual. He wrote to Charles Bunnell, president of the University of Alaska: "No doubt they at the Department think I'm a bit naive, anyway, to come in and apply for a governorship without even a single politician to back me up and offer to a Republican administration the names of three died-in-the-wool Democrats as my principal character references. I gathered that in this Alaska gubernatorial race some weird specimens showed up. They probably considered me the strangest of the lot." (Before Dunkle formally submitted his name, he had received the endorsement of Robert J. McKanna, the Republican national committeeman for Alaska and a regular correspondent of Billie.) Dunkle's name made the list for governor very late. The position was soon filled with a man who was well regarded by Dunkle: Frank Heintzleman, the longtime regional forester for Alaska. Dunkle wrote to Andrew Nerland, "I found that the matter had just about been settled and that the choice was for a man whom I consider very satisfactory."[52]

Special Assistant Secretary Raymond Davis in the U.S. Department of the Interior then told Dunkle that he should apply for the presidency of the University of Alaska. Dunkle reminded Nerland, who was chairman of the university's board of regents, of his qualifications:

> As you know, I have been actively at work up there for many years. . . . [T]his work, although I feel that it has contributed a lot to the development of Alaska, has been almost entirely in private industry, and I would like to round it off in something directly concerned with the public welfare of the Territory. . . . [M]y work has led me to deal . . . with men in all walks of life including financiers such as J. P. Morgan, Jr. and the Guggenheim brothers, executives, managers, engineers, workmen, stockholders and others connected with some of the largest mining corporations in the world. . . . I have more or less close acquaintance with corporations and individuals who have taken considerable money out of Alaska and perhaps I could persuade some of them to put some of it back as needed by the University.[53]

Dunkle then cited his own educational background and extracurricular strengths. He also wrote to Charles Bunnell, the first president of the University of Alaska, to follow up on his formal application through Nerland. Recognizing that his age (sixty-six) might be an issue, Dunkle thought that his ability to work twelve hours a day seven days a week should be proof of something. "As to mental processes, I do not notice any great change. I seem to be as good as I ever was."[54]

Age did turn out to be a factor, as did Dunkle's lack of advanced degrees. The executive committee of the regents decided that it would search for a candidate between thirty-five and fifty-five years of age, and the Alaska legislature was considering degree qualifications for the position. Dunkle's immediate reaction was to suggest that Billie should apply; she was within the age window and had a doctorate. The regents, however, decided to limit their search to men.[55]

Although Dunkle failed to win either appointment, he had another idea that approached public service. He resolved to record and write Alaska's history.

12

One More Experiment

The worst of the Dunkles' financial battles were over by 1953. Mining debts had been resolved. Billie was beginning to be known as a power in Washington, D.C. The resultant financial stability allowed Earl to work on a challenging new project, a history of his life and times in Alaska. To assess its potential value, Dunkle shared his idea with another authority on Alaska's Mt. McKinley region, Bradford Washburn of the Museum of Science in Boston. Dunkle wrote to Washburn:

> My work there [in Alaska] took me to all parts of the Territory by all means of travel and I became well acquainted with the geography and geology of the country as well as the people in it.
>
> For one example take the Kennecott mine. The story of this mine should command general interest because it was the richest copper mine ever found in the world and its discovery, development and operation make a story of exploration and business romance which is, in my opinion, hard to beat. It reached its culmination in the formation of the largest copper corporation in the world. Its story is the pre-eminent one of all that took place in the exploration and development of the Copper River region, but it is only one of many. I numbered amongst my best friends the men who found it, owned and mined it. . . . A proper history or story . . . has not been written of that section of Alaska which lies south of the Alaska Range and across the full breadth of the country, which is the region with which I am most familiar. My life there and that of my friends and acquaintances ranges through the early pioneer conditions of living and working up to those of the present. I believe such a story should be written. . . . The old timers are nearly all gone and it will not be long before the opportunity to get a lot of this at first hand will be lost.[1]

Dunkle thought that, with proper help in assembling and editing, he could write the history. He asked Washburn for his opinion and for publishing suggestions. Income was still an issue for the Dunkles because Earl wrote to Washburn, "One of the main ideas, of course, in addition to getting the people and events into a record, is to try to make some money out of the thing."[2]

Dunkle's letter of inquiry was written only a few days after he had written a long venture-recounting letter to Washburn, who answered Dunkle immediately: "I cannot tell you how much I appreciate the wonderful job which you did in answer to my letter. It was extremely interesting to re-read the story of that dramatic trip of yours across Anderson Pass but the many other experiences and anecdotes about early days in the McKinley Park region are fascinating to read—particularly for someone as interested as I am in that part of the world."[3]

Washburn agreed that Dunkle's story should be written, but he cautioned him that there was a limited readership for books on Alaska. Nevertheless, Washburn sent the two long letters that he had received from Dunkle to his friend Charlie Blanchard at the Boston publishing house of Little, Brown and Company. Blanchard took them and his own favorable opinion to Howard Cady, the editor in chief. Cady's reaction, as expressed to Washburn, was also favorable: "I agree with you that someone should be persuaded to work with Mr. Dunkle and put together a book based on his experiences. It could be one of those 'as told to' or 'in collaboration with' arrangements and, if there is enough good material, I would guess that a book of that sort might have quite a satisfactory sale."[4]

Dunkle's history project actually began in January 1953. Earl; his Kennecott-era friend Henry Watkins, who was then living in nearby Virginia; Henry's wife, Ruth; and Bernie McDonnell, Billie's new associate, met at Billie's home to discuss and record Alaskan history. The neophyte historians had maps, photographs, and correspondence to review and, for recording, a brand new reel-to-reel tape machine. Henry Watkins's reminiscences were critical to the project. He had arrived in Alaska in January 1900, ten years before Dunkle, and he had an excellent memory. A near contemporary of Watkins, L. A. Levensaler, could not attend the recording sessions, but he had written Dunkle a long letter about his own considerable part in the early Kennecott venture. Moreover, thanks to the forced circumstances of World War II, Levensaler and Watkins had been in regular communication for many years. Levensaler had gone back to Washington in 1942 for what was supposed to have been a short war-materials job, but he ended up staying until 1946. During much of this interval, Levensaler spent leisurely Sunday breakfasts at the Watkinses' home, exchanging memories of Kennecott and the Wrangell Mountains.[5]

The tape-recording enterprise extended over several months. The final product, with some overlap and repetition, is about fifty hours long. Kennecott and its personalities are the main topics of the first reels. Watkins and Dunkle told anecdotes, described from firsthand knowledge how to live and travel with relative safety in the Alaskan bush, and contrasted the costs of 1910 with those of 1953 Alaska. They remembered which roadhouses were the best or the worst and what they were charged for a meal or a night's rest in a crowded bed. The recordings describe the rise of commercial aviation as well as Alaska Steamship Company, its officers, and operations. Dunkle thought that, for a period, Alaska Steam had equaled any steamship company in the world in operating efficiency. But Dunkle also knew (and tried to persuade Bernie McDonnell, who sometimes questioned the more informed participants) that Alaska Steam's heyday had been superseded by commercial aviation and by the heavy trucks then beginning to traverse the Alaska Highway through Canada. The growing importance of tourism to Alaska was noted, as were two early visitors to the Great Land: Capt. James Cook and Jean-François de La Pérouse, explorers who had sailed to Alaskan shores in the late eighteenth century.

Dunkle and Watson told intriguing stories about aviators, prospectors, con men, musicians, prostitutes, and other Alaskan characters. Their accounts suggest remnant lodes for future historians: Henry Watkins; Willis Nowell, classical violinist turned prospector; mining raconteur Jack Baragwanath; adventuresome Helen Green Van Campen; and Watkins's contemporary, Lawrence A. Levensaler. Dunkle probably would have recounted stories about these fascinating individuals in a rough manuscript that he had promised editor Howard Cady, but first, he had to complete some work in Alaska.[6]

The engineer in Dunkle could not be held down. Experimentation to improve the thermal quality of a low-rank coal at Broad Pass, Alaska, displaced Dunkle's history project. There was an economic implication as well. As had World War II, the Korean War, with its consequent expansion of military bases in Alaska, made coal more valuable and stretched the resources of existing mines. Dunkle knew of three coalfields near Colorado Station that might supplement the existing sources of the fuel as well as augment the Dunkle family income. The first was the coalfield at Costello Creek, where Dunkle had mined during and after World War II. The second was a thick seam of coal exposed in the canyon of the

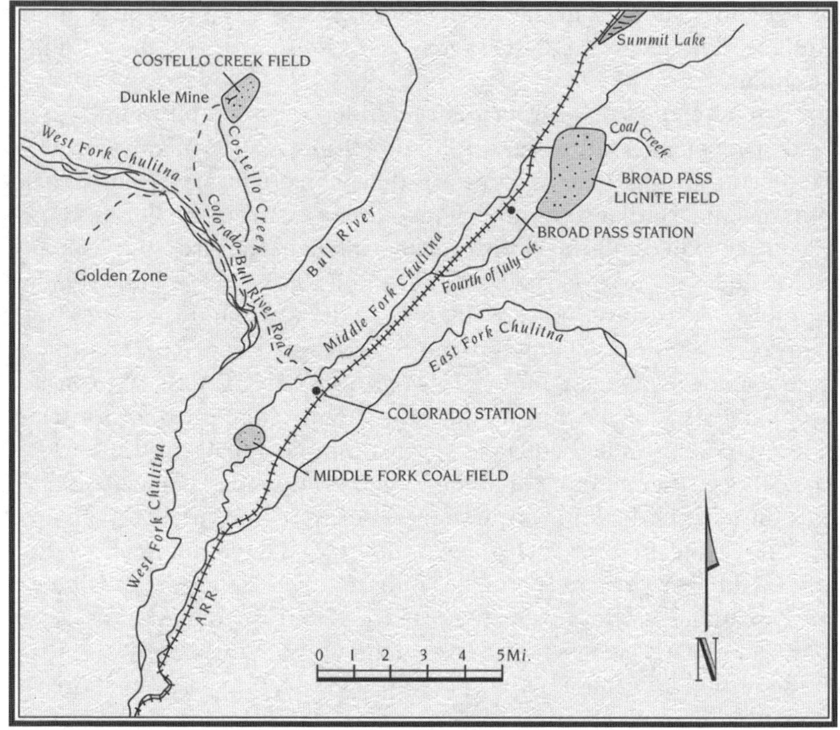

Figure 39. Coal deposits near Broad Pass and Colorado Station. Dunkle developed the lignite deposit at Broad Pass. Author's compilation.

Middle Fork of the Chulitna River. The third was a coal outcrop within a few hundred feet of the Alaska Railroad, three miles north of Colorado Station near Broad Pass (fig. 39).

Dunkle discounted the Costello Creek field because Ken Hinchey of Anchorage had open-pit mined most of the resource that had remained from Dunkle's underground operation. The other two fields, especially Broad Pass, had potential. In addition to being accessible, the coal beds at Broad Pass, Dunkle believed, were almost flat and barely covered by glacial gravel; hence, they were easily mineable. The size of the field was uncertain, but in 1944 G. A. Apell of the U.S. Bureau of Mines had said that the inferred tonnage of coal was "unlimited." Although they recognized the tonnage potential at Broad Pass, Apell and most other government engineers discounted the commercial potential of the field because the coal was lignite, a soft coal with relatively low heating value. Dunkle thought that he could increase the heating value of the lignite, and thus increase its economic worth.[7]

Several properties determine the commercial value of a coal. Heat value (content), which is expressed in either British (Btu) or metric (caloric) units, is the main parameter. The amount of ash and moisture, the temperature of ash fusion, trace elements, and sulfur content are secondary but often critical factors. Dunkle and government geologists and engineers ranked the Broad Pass coal as a high lignite with a heat content of almost 7,000 Btus per pound, about 1,000 Btus less than the coal in the nearby subbituminous fields.

Moisture is a variable that, at least in theory, can be manipulated to increase the heating value of coal. Small amounts of water are present on the surface and in the cracks of the coal, but the more important and obstinate portion is the moisture held within the pores themselves. This part is sometimes called *equilibrium moisture.* Tightly compressed anthracite has the least equilibrium moisture: a few percent or less. The lowest rank coal, lignite, commonly contains at least 30 percent. If a coal is dried at high temperatures, the equilibrium moisture is driven off, but it is largely resorbed over time as the coal is cooled and exposed to the atmosphere.

Dunkle was one of the first to tackle coal enhancement quantitatively. He reasoned that reducing the moisture content would increase the thermal value of the coal as well as decrease its shipping cost. The technical part of the problem is fairly difficult. The coal has to be dried but in such a way that its fabric and inherent strength are not destroyed. Once dried, the coal must be treated to stop resorption of equilibrium moisture. Another difficult aspect is economic. The cost of drying must not exceed the benefit; otherwise it is like designing a perpetual motion machine: intriguing but illusory.

Dunkle's solution to the first aspect of the technical problem was to dry the coal under one atmosphere of steam pressure, essentially at one hundred degrees centigrade. Under these conditions, about two-thirds of the moisture was eliminated. Broad Pass lignite, originally with about 30 percent moisture, was converted into a new product with 10 percent moisture and about one-third greater heat content. The structural fabric of the coal was preserved. Doug Colp, an engineer who visited Dunkle's project at least twice, recalls that the dried coal resorbed moisture rapidly, so the improvement was only temporary.[8] Dunkle solved that problem by treating the dried coal with fuel oil, which apparently filled the near-surface pores of the coal. Oiling also added a small amount to the heat content of the product.

Dunkle recognized the cost factor and weighed it against the economic benefits. He could not resist trying to explain the project, even to a probably disinterested party. Dunkle told Mrs. Ferrick, his son Jack's mother-in-law, "My experiments show that the water can be removed at a cost less than the saving in the freight . . . , and after it is removed the coal becomes a superior rather than an inferior fuel."[9]

The Broad Pass project began in 1954 and extended into the early 1960s. Dunkle underestimated the length of time necessary to complete it. The year 1954 was mostly consumed by acquiring prospective coal land from the U.S. Bureau of Land Management, establishing a prospecting program, and forming a corporation to back the project. He used the rest of 1954 and most of 1955 to explore both the Middle Fork and Broad Pass fields to determine whether there was sufficient coal for further work.

Dunkle tackled the Middle Fork coal first. Expenditures, mainly for about four miles of new road, totaled $5,790 through 31 December 1954. The costs included the rental of William A. ("Bill") Waugaman's bulldozer and the operator's time. This is when, as recounted earlier, Waugaman, who was strong and fit at thirty-something years, could hardly keep up with sixty-seven-year-old Dunkle. It is also when Dunkle, recalls Waugaman, pointed out which holes in the roof of their primitive cabin at Colorado Station were due to normal wear and tear and which were made by a grizzly bear. Although the coal seam in the Middle Fork was thick, it dipped steeply, and the potential mineable tonnage was limited. Dunkle thereafter concentrated on the Broad Pass field.[10]

Drilling in Broad Pass confirmed Dunkle's model of mineable flat-lying coal and suggested that there were at least eight million tons of it in three seams. There were other possible (inferred) reserves. A post-Dunkle report of 1961 projected a potential resource of more than 100 million tons for the Broad Pass lignite. Drilling results were so good that Dunkle, miner-promoter Keith Capper, Brown, and Wendell Barnes, while in a merry mood, concealed the *P* in *Broad Pass* before they posed for a snapshot next to the sign at the railroad station.[11]

Two tasks remained: first, test mining to obtain large samples on the order of hundreds of tons; and second, testing to determine if the coal quality could be enhanced economically. The first bulk samples were collected to determine if the lignite could qualify for sale on an as-mined basis. As Dunkle explained to his son Bruce, "Our job for the winter [1955–1956] is to drive a tunnel for a hundred feet or more into the Dunkle seam (14 feet thick) and to ship out average samples . . . to determine the efficiency with which it can be burned in a boiler, so that we can offer a bid to sell coal to the Army or CEA [Chugach Electric Association of Anchorage]."[12] The test results were essentially a foregone conclusion. Lignite, as mined, could not compete with the subbituminous and bituminous coal then being produced by other operators in Alaska.

Dunkle believed that his plans to enhance the fuel would make it competitive. For experimental purposes, he designed a one-ton reactor, where the lignite could be subjected to the requisite temperature and atmospheric steam pressure. Mining of the coal sample was completed in late October 1956. After the miners had completed their work, they left Dunkle and quite a few tons of coal at Colorado Station. In one letter to Bruce, his father made the job seem very simple, although the weather was adverse: "[A]ll I have to do is prepare a couple of carloads of dried coal for tests, and then there will not be a great deal for me to do here. . . . I could stand a little bit of soft weather back there. . . . October was the coldest October on record, the average at Fairbanks for the full month being more than ten degrees below any previous average, and that is a lot."[13]

The weather did not break. During the first week of January, a full-scale blizzard dumped so much snow that Dunkle's Quonset building at Broad Pass collapsed. Even with good weather, the job was almost impossible for one man. The coal was outside and covered by several feet of snow. It had to be uncovered, shoveled into the reactor, dried, treated with fuel oil, and then shoveled back outside. Processing one ton of coal was equivalent to several tons of work. Moreover, it took several hours a day just to provide living necessities. Dunkle's camp was so minimal that Peter Bagoy Sr., who visited there in September 1957, described it as "worse than Siwash." This description, coming from sourdough Bagoy, could only mean a very poor camp. In addition to cold weather, heavy snow, and lack of help, the project was hindered by a lack of equipment. A new steam boiler and motor for the prototype batch drier had not arrived by 26 January 1957, although they had been ordered several months before.[14]

Financing, or the lack of it, continued to be a problem in the winter of 1956–1957. David Foote, a business associate of both Earl and Billie, pulled out of the project in the fall of 1955, just as it was getting underway. Bertram C. Ricks, who had helped Earl organize Alaska Exploration and Development Company to fund the project, was unsuccessful in his financing efforts. Another man, Keith Capper, a Republican office seeker, miner, and promoter, was little better at delivering the funds when they were needed by the project. There were also needs to be met in the geographically extended Dunkle family, needs that helped Dunkle keep his problems in perspective.[15]

Billie Dunkle, who was still in the east, helped out whenever she could, but she had less and less free time during the period of Earl's self-imposed hardships in the Broad Pass field. Billie and her associate Bernie McDonnell had been active in Eisenhower election campaigns since 1952. The 1956 Eisenhower campaign was accelerating. Billie was moved from Washington, D.C. to a new position as deputy director of Women's Finance for the campaign, which was headquartered in New York. She was even busier than before. During this time, Bruce Dunkle was often the bond that tied his parents' long-distance marriage together. Both parents encouraged Bruce in his own endeavors, but they also shared with him their activities. Bruce, as an heir and future manager of the Dunkle mining ventures, was continually briefed on the projects.

In the spring of 1955, Bruce Dunkle passed the Annapolis eligibility test and enrolled in the Fleet Academy Prep School in Bainbridge, Maryland. A few months later, he passed the test for fleet admission to Annapolis and entered the midshipman program. Bruce's Annapolis appointment played a significant role in all Dunkle correspondence during the next two years, more or less forming a mean between two strongly contrasting lifestyles. Bruce had sparsely furnished but warm Navy-neat housing with all the food he could eat. At Colorado Station, Earl had quarters barely better than a tent. Billie's living accommodations were much better than Bruce's spartan ones at Annapolis and almost infinitely better than Earl's quarters at Colorado Station. When Bruce entered Annapolis, Billie looked for a still nicer home. As she explained to Bruce, "I am dickering for a really super place only a few miles from Annapolis. Three bedrooms, three baths, deep freeze, screened porches, fifteen acres, lovely water, two boats, pier, pavilion, bath house, two-car garage. . . . Dad would be delighted with the work shop."[16]

Correspondence kept the family close, and even in Alaska, radio and newspapers quickly reported current events. Father and son could gloat, commiserate, or fume at Navy's or Yale's athletic results. In 1956 Navy had an especially good football season, but Yale lost to Colgate before barely beating Princeton. Navy went on to beat Army in one of the legendary battles between the two teams. The narrow victory, which was solidified when a midshipman fell on an Army fumble, triggered further details from Earl about the ancient athletic happenings at Yale, where unanticipated events had also changed almost certain endings. Bruce probably knew the stories by heart.[17]

There were also contrasts in the social lives of the Dunkle family members. At least in his plebe year, Bruce's life was tightly constrained. As deputy director of Women's Finance for the Eisenhower campaign,

Billie had a busy political and social schedule that at times included the finance director, Helen Hayes (fig. 40). Billie's schedule also included her duties as secretary of the Washington State and Alaska Society and secretary of the Washington Home Rule Committee.[18] Earl visited friends and attended meetings of the Quiet Birdmen when he was in Anchorage, but at Colorado Station his life was solitary and his entertainments were simple. Some, which he enjoyed greatly, came from a wilder kind of company that often furnished him with filler material for his letters. He especially liked to watch the avian antics of the Canada Jays, the camp robbers that he fed every day. There was also a lone woodpecker that was apparently saved by species from the harassment that the jays meted out to their own kind. The camp robbers would swoop down on their food, but not so the woodpecker. Father wrote son, "I get a laugh out of him. If a piece of the rind drops to the surface of the snow he doesn't fly down to get it but backs down the pole, digs out the piece and hops all the way back up again." The woodpecker was not above playing peekaboo with Dunkle. Earl reminded Bruce that the woodpecker's tongue was sharp and pointed "like a little spear." A rare visitor, the first that Earl had ever seen close up in almost fifty years, was a wolverine, which raced along the ruts in the snow-packed roads at Colorado Station and refused to be intimidated by the vehicle that Dunkle was driving. As Dunkle described it, "He made a quick jump to the side and whirled as he did so ending up facing the car and all still full of fight . . . a unique and rare experience." It was hardly news when a moose did almost the same thing, following the path of least resistance up snow-plowed railway tracks. News was when a moose actually escaped the train. Earl tried to shoo one off the tracks at Colorado Station, but the moose stayed in front of the train to the Broad Pass section, where it finally escaped and, exhausted, posed for pictures taken by the railway section boss, Ernie Kirsch.[19]

Occasional Friday evenings with the Kirsch family at the Broad Pass railway section house were probably the high points of Dunkle's long winter of 1956–1957. Contacts with Kirsch or his section hands, Charlie Gore and Laddie Kavoda, were common. All of the railroaders looked out for the seventy-year-old Dunkle at Colorado Station. On Friday evenings the men, a few neighbors or visitors, perhaps "Walking" Swazey, who sold insurance up and down the line, and the Kirsch family gathered for supper and entertainment, usually the sixteen-millimeter sound films furnished by the railroad. Dunkle was a frequent guest on Friday nights; he took a shower and visited with the Kirsches. The Kirsch children were a special pleasure. Earl made young Johnny Kirsch a windup

Figure 40. Billie Dunkle (top) *in her role as deputy director, Women's Finance, Citizens for Eisenhower-Nixon, 1956. Helen Hayes, the director, is at lower left. Publicity photograph. Courtesy, Dunkle family.*

floorwalker out of spools and carved a recorderlike wooden whistle, probably reflecting skills and toys from his own boyhood. Later Dunkle sent Rose Kirsch a huge pink beach towel, which kept five Kirsch kids fluffed up for years.[20]

In Washington, D.C., Billie's schedule relaxed after Ike's inauguration in the winter of 1957. As a presidential campaigner, Billie was now a lame duck, but it was a good time to convert her staff work into patronage. She found a position in the U.S. Department of the Interior as the director of a new program on native arts and crafts in Alaska. The position, a temporary one with possible extension, seemed tailored for Billie's knowledge of Alaskan geography, her artistic talents, and her administrative abilities. The Department of the Interior's representative told Billie, "You as liaison officer will be the key" to the program. Billie's ideas for the project, which were probably outlined for a staff presentation, were global, ranging from jade to blubber to on-the-job training. She also noted the importance of tourists, Nome Skin Sewers, horns and hooves, and rammed earth construction in the far north. Her approach was often political. She wrote to new territorial governor Mike Stepovich, "As I warned you at the picnic at Lee McKinley's farm, I plan to make use of that open door of yours. At present I am hard at work on an Alaska Pilot Project, summary enclosed, which I hope will have your interest and approval."[21]

Earl wrote Bruce that the arts and crafts program was an important one for Alaska. It also had a tangible personal benefit because it occasionally brought Billie to Alaska. Shortly after her appointment, Billie visited Anchorage and Colorado Station, and both Dunkles had an opportunity for relaxation and conversation. By this time, the coal dryer was finally completed. Earl had also found a technically sound partner for the project, Goodnews Bay Mining Company. Goodnews Bay had a mine in Alaska and a headquarters in Seattle. The company mined platinum instead of coal. It was led, however, by Andrew and Edward Olson, who had always been innovators in Alaskan mining. In contrast to Dunkle's earlier partners in the coal project, Goodnews Bay was financially solid, and the Olsons' associate in the company, Charley Johnston, was a good administrator and contact for the Dunkles.[22]

In mid-September 1957, Billie returned to Washington, D.C. for a meeting of the Indian Arts and Crafts Board, where she was scheduled to open the afternoon session on 27 September. Although the Alaskan village program was only a few months old, Billie believed that considerable progress had been made and that it must be protected. She told the assembled group, "The wide support and co-operation which have been pledged is indeed a step forward in view of the Territorial agencies' tug-of-war history. To say that this great advance is an operation completed for all time would be in error. We have made what is justly recognized as a wonderful beginning."[23]

Earl took Billie to the airport in Anchorage for her return trip to Washington, D.C. Before returning to Colorado Station, he found a letter from Bruce, which he answered seriously, but he also could not resist making one dig: "Excuse me for mentioning it, but I hope your prophesy as to Navy's football chances proves to be better than your baseball choices last spring." A few days later, Earl was back in Anchorage. He sent another quick note to Bruce before he caught the train for Colorado Station. He reported that the coal project at last was healthy; as he told Bruce, "The financial outlook for the mine has greatly improved, which is quite a relief."[24]

Dunkle needed one more ingredient for the coal plant—a clear-water spring—and he had a good idea where to find one. He visited briefly with the men at the Broad Pass section house on the morning of 29 September before heading east on foot toward Fourth of July Creek, his chosen plant location. When he left Broad Pass, the weather was clear. As the skies lightened, however, there was a hint of the high, wispy ice clouds that foretell snow or rain in the Alaska Range with almost perfect accuracy. Dunkle did not return to Broad Pass on the evening of the twenty-ninth as expected, or on the thirtieth. Section boss Ernie Kirsch sent word up and down the railroad. A search was started immediately, but a cold snap sent temperatures to near zero and dropped as much as eight inches of snow. Word of Earl's absence was received in Anchorage on 2 October; by the fourth, Billie had been alerted, and dozens of civilian volunteers were moving toward Broad Pass. In Anchorage Dan Cuddy had to make only one telephone call to Ft. Richardson before soldiers of the Twenty-third Infantry boarded a train for the pass. Pilots and close friends Jack Carr, Al Lindemuth, Harry Bowman, and Gren Collins were airborne on the morning of the fourth. Bowman and Collins, with perhaps more knowledge of Dunkle's plans than either Carr or Lindemuth, landed and began to search on foot. They found Dunkle's body partly covered by snow. After seven decades, his well-worn heart had finally failed. The troops, who had waited for word from Bowman and Collins before they joined the search, brought Dunkle's body back to the railroad. Billie and Bill Dunkle arrived in Anchorage on 5 October to prepare for the funeral at the Elks Club. Earl was survived by his wife, his three sons, and his sister, Murna, who was then living in Boston.[25]

The pallbearers were longtime friends from Anchorage: Harry Bowman, Jack Carr, Ray Peterson, Bob Romig, and Lawlor Seeley. Bowman, Carr, and Peterson were from Dunkle's aviation world; all of them were skilled outdoorsmen. Medical doctor Romig and dentist Seeley had been friends and shareholders in the Golden Zone mine since before World War II. Dunkle left many friends worldwide in the mining industry. One of them, DeWitt Smith, shared his thoughts with Billie: "Earl enjoyed life and the tougher the better. I am sure that he would have chosen to die in harness in the field had he had his choice." Smith was Dunkle's Yale classmate. Earl had planned to attend the fiftieth reunion of the 1908 Sheffield mining class the following June. Smith told Billie that Earl "will be sadly missed there." Another of Dunkle's classmates, Howard Oliver, emphasized both of Earl's careers: "His . . . adventures in flying over the mountains and wilderness of Alaska were as daring and thrilling as his conquest of mother earth by the mining he learned at Yale."[26]

Billie Dunkle soon abandoned the position that she had tailored in the Department of the Interior so that she could continue Earl's Alaskan projects. As with every other project that she attempted, Billie threw herself into mining development. Earl's partners, the experienced men of Goodnews Bay Mining Company, were willing and financially able to help. The coal-drying experiment had consumed months with Earl trying to do everything. Within days Billie, after consulting Charley Johnston of Goodnews Bay, had arranged for the batch coal-plant test to go forward under the direction of engineer James March. March, with the help of Rocky Cummings, Jim Beaver, and Eddie Barge, who were pioneers from the nearby village of Talkeetna, processed thirty tons of Broad Pass coal, one ton at a time, in Dunkle's batch reactor. The coal, which had been reduced in weight to twenty-one tons by drying, was tested at the Fairbanks Exploration Company coal power plant in Fairbanks. The results were promising. As predicted by Dunkle, the moisture was down by about two-thirds and the thermal value was up even higher than the run-of-mine subbituminous coal from the Healy fields. Later cost analysis showed an apparent benefit for Broad Pass dried lignite against most other coals. Although dried Broad Pass lignite would cost more than Healy River coal on a per-ton basis, it would cost less than Matanuska coal. Basing the cost on the actual thermal value, the Broad Pass product was the more economical coal, although in some cases the margin was slight. Based on even a slight margin and the policy of the military and utility plants to expand their fuel sources, processed Broad Pass lignite could compete.[27]

Billie used her business acumen to move the project forward. More coal land was leased, knowledgeable and friendly associates were found, and a well-known coal-plant engineering firm, McNally Pittsburg Manufacturing Company of Pittsburg, Kansas, was engaged to convert the batch process into a continuous mechanized one. Billie formed and presided over a new company, Accolade Mines, Inc., to finance the project. In a prospectus sent to the Alaska Natural Resources Department commissioner, Phil Holdsworth, Billie proved herself to be a good mine promoter, who was equally at home with the financing and the technical sides of the operation. She reported that McNally Pittsburg had made progress in designing continuous processors and was pursuing both Dunkle's steam equilibrium drying and an infrared variant. Both methods appeared to have commercial promise, and Billie was making some marketing progress.[28]

Billie did not neglect other opportunities or obligations, especially those concerning the Golden Zone property. Billie renamed the property to reflect her assessment of its value and to escape the negative connotations of GZM, which by then was remembered as another ill-fated mining dream. Billie thought that the chief Roman god would be an appropriate symbol for the mine; thus the Golden Zone became the Alaska Jupiter. Billie also did something that Earl should have done fifty years before: She joined the AIME, the professional society for mining and metallurgical engineers.[29]

Always a pragmatist in matters of income, Billie looked for jobs that would not require the constant work and travel of the Indian Arts and Crafts Board. One opportunity that she took was as a consultant to the federal Small Business Administration, a position that gave her more time for her other interests. In the fall of 1961, Billie had surgery and never recovered completely. She died on 27 April 1962 in Washington, D.C.[30] Her life was as eventful and pioneering in its own way as that of her husband. She was a scientist, inventor, artist, author, mother, and investor and promoter in both mining and aviation. In postwar Anchorage Billie was a scrap dealer, but earlier she had been that city's social doyenne. She had also been socially prominent in Scotland, South Africa, the Pacific Northwest, and Washington, D.C.[31] It was the end of a world-spanning odyssey for a woman of somewhat uncertain origin, who grew up in the borderland hills of southern Scotland.

It is a mining dictum that nature makes mineral deposits but men make mines. Billie Dunkle was an independent woman who complimented the mining career of her husband by accepting his goals and making them hers. Together the Dunkles made a great team.

Epilogue

THE LAST GREAT GOLD RUSH IN NORTH AMERICA occurred between 1896 and 1910. Four stampedes—to the Klondike in 1898, to Nome in 1899, to Fairbanks in 1903, and to the Iditarod in 1910—coalesce, in the larger view, into one event. Scholarly work continues to add new facets to the history of the gold rush, but it is ground well turned. Already it blurs into legend. In 1953 Earl Dunkle proposed to tell a story that overlapped the gold rush. It began with the discovery of the Bonanza copper lode in the Wrangell Mountains in July 1900 and continued through midcentury. Dunkle's story encompassed copper, gold, aviation, and other aspects of a developing Alaska. He gathered materials and made a significant start on the project but did not complete his task.

As then envisaged by Howard Cady of the publishing house of Little, Brown and Company in Boston, Dunkle's story would have been told with the help of an experienced writer. It almost certainly would have stressed the danger and excitement of Dunkle's life, the richness of the copper bonanza, and perhaps his experiences with Stephen Birch and the oft described Guggenheims. Like this account, it probably would have emphasized Dunkle's Yale connections: the tale of a young Ivy League geologist and engineer who came northwest to adventure. Perhaps such a story would have restored Dunkle's earlier fame and fortune. However, it probably would not have encompassed the scope of Dunkle's interest in the history of Alaska Steamship Company. It would not have had the benefit of time to fully assess the development of commercial aviation in Alaska. And it probably would have minimized Dunkle's work with coal.

Perhaps it is better that Dunkle's story has been deferred so history can compare the gold and copper rushes with the early conservation movement in Prince William Sound and the Wrangell Mountains of Alaska. In several respects, the Copper River story that Dunkle proposed seems more significant to Alaska, the Pacific Northwest, and the nation than does the more glamorous gold rush. The gold rush brought more people to Alaska than did the copper discoveries, but many of those visitors left quickly. It enticed a few million dollars from the Guggenheims for gold-mining projects in the Klondike and at Flat, Alaska, but their interest was not sustained by the yellow metal. In contrast, the discoveries of rich copper deposits in Prince William Sound and the Wrangell Mountains justified decades-long investment in Alaska by the Alaska Syndicate and its corporate successor, Kennecott Copper Corporation. That investment created both copper mines and an international steamship line, which survived until the advent of container shipping. The Copper River tested the best that railroaders and bridge builders had to offer. The completed railway was the first significant link between the coast of Alaska and its interior.

The copper stampede was also more significant than the gold rush to the emerging conservation movement. The gold rush was fomented by individuals, who made elusive targets. The "Guggenmorgans" of Copper River, however, made bigger and more vulnerable targets. It is arguable whether Theodore Roosevelt's withdrawal of coal lands from entry in Alaska was justified, but the result was a major shift in federal resource policy toward conservation and collectivism.

For a time at the beginning of the twentieth century, Alaska was as important nationally for its copper and coal as it is currently for its oil and natural gas. That combination of copper and coal ignited a political wildfire in Alaska that swept beyond the boundaries of the territory. No one has summarized this story as well as James Wickersham did in 1923:

> The struggle for home rule and an American type of government in Alaska has now continued for more than half a century of American domination. It is an historic struggle; it brought on the Pinchot-Ballinger investigation in Congress; destroyed the friendship between Theodore Roosevelt and President Taft; split the Republican party into two great factions; defeated President Taft for re-election in 1912; elected Woodrow Wilson President of the United States; and changed the course of the history of our country.[1]

It is still difficult to resolve the resource value of the Bering River coalfield, the field near Katalla on Prince William Sound that triggered Theodore Roosevelt's withdrawal of coal lands. Earl Dunkle told Stephen Birch that the coalfield was not viable because of its complex geologic structure and because the coal's ash did not cake properly in marine boilers. The best chance the field had for development was in the early 1900s, when hundreds of workers were poised to mine coal for the Guggenheims. Their opportunity was denied by questionable charges that stopped coal mining during the critical stage of copper-mine development (1907–1911).[2] Did Dunkle consider the implications of changed coal policy in his assessment of the coalfield for Birch? The answer is uncertain. Dunkle knew the coal miners, J. P. Morgan Jr., and the Guggenheims. He probably would have been the best person to ask because he also knew the geologic and engineering characteristics of the field.

The perspective of time and acquired knowledge allows evaluation of other Dunkle projects. One sentence in his 1954 paper on the Copper River district explains why he undertook so many: "However, in a new country with no past mining experience to serve as a guide, a sure knowledge on which to base such interpretations can be gained only by the expenditure of time and money in development work."[3] In essence all of Alaska was new, and Dunkle thrived on the new and the innovative. His ideas and skills contributed to the successful operation of the Kennecott mines in Alaska, established the Lucky Shot as the premier mine in the Willow Creek district, and furthered Alaskan aviation. He helped to create the seaplane port at Lakes Hood and Spenard in Anchorage when it was most needed. A successful aviation company, Alaska Airlines, has emerged from Star Air Service, which Dunkle and his associates established seventy years ago. Other Dunkle projects deserve further testing. The unique low-head design of the hydroelectric system at the Golden Zone site—with no dam and no permanent effect on any fishery—might be adapted for other broad glacial rivers in Alaska. The Broad Pass coalfield remains as the Dunkles had left it. Dunkle's work on the coal dryer was innovative. His drying method has been enhanced and now offers an opportunity for developing Alaska's lignite deposits.

A few projects cannot be tested. The Dunkle coal mine at Costello Creek, the Shannon deposit near Slippery Creek, and the Caribou Creek placer deposit are in Denali National Park and Preserve, a location that permits no substantial mining operations. Arguments about undiscovered copper deposits in the Wrangell Mountains are moot because the Kennicott-McCarthy area lies within the Wrangell–St. Elias National Park and Preserve. In June 1998 Kennecott Copper Corporation donated its last mineral rights in the region to the National Park Service, completing a transaction first contemplated in 1938.

At the Golden Zone mine, Dunkle's ore estimates were proven to be conservative by more drilling and tunneling, although the mine has not produced ore since 1942. Production would require the incentive of an increased gold price. Dunkle and others of the old gold-mining fraternity believed that at some time gold would be revalued. They had little faith in the long-term worth of paper money.[4] In the mid-1960s, intense pressure on the gold held in reserve by the United States and valued at thirty-five dollars per ounce reached a breaking point. The country would either have to find rich new deposits that could be mined at the thirty-five-dollar price, or the price would have to give.[5] Despite discoveries of rich gold deposits in Nevada, the price broke, and the present two-tier system for gold was adopted in 1968. Dunkle's view on the long-term worth of gold was vindicated.

Earl Dunkle's life in Alaska spanned a large part of its early history under the U.S. flag: from forty-three years after its purchase until one year before its statehood. It spanned a smaller but still significant interval of national history. The worldwide promise of the "Good Years" and La Belle Époque was broken by World War I. Nevertheless, Dunkle and his 1908 classmates from Sheffield never lost the can-do, must-do ethic of that first decade of the twentieth century.

Earl Dunkle loved two intelligent and independent women: Florence Hull and Billie Rimer. Although Florence is concealed by time and overshadowed by the more public Billie, Florence's sons, Bill and John Dunkle, would not have traded her for any other mother. Florence did not marry a governor, as her Oberlin classmates had projected. She married a magician, who turned mineral deposits into mines. At times the realities of inflation, shifting ground, or a world war showed that magic is, after all, illusion. Dunkle—and his life partners, Florence and Billie—did not, however, accept defeat without a struggle.

Three people—Earl Dunkle, Florence Hull, and Billie Rimer—came to Alaska with little more than dreams in their pockets. They lived those dreams according to the advice that Jack Baragwanath's wife, Neysa McMein, received from her father: "[O]nce when I was quite a little girl my father and I were going to the corner in Quincy to take a streetcar. Before we got there we saw the trolley coming and he grabbed my hand and said, 'Come on, let's run!' I said, 'We'll never make it—we'll miss it.' 'All right,' he said, running faster, 'let's miss it, trying!'"[6]

Notes

In citing works in the notes, short titles have generally been used. Complete citations are listed in the bibliography. Some notes include abbreviations from the list at the beginning of the book.

ACKNOWLEDGMENTS

1. Simpson, "Bonanza–Motherlode Mine," 7–9; Simpson, *Legacy of the Chief.*
2. Green, *Gold Hustlers.*
3. Stoll, *Hunting for Gold.*

CHAPTER 1: HERITAGE AND EDUCATION

The source of most of the information about the heritage and family life of the Dunkles is a genealogical typescript compiled and written in 1977 by Clara Adella Dresskell Page: " 'Dunkle'-Dusk-Twilight-Evening." In addition to genealogical charts, Page's manuscript includes newspaper clippings on various members of the Dunkle family. The manuscript repeats, verbatim, accounts that Peter S. Dunkle wrote ca. 1926. Peter Dunkle was Earl Dunkle's uncle. In this chapter, short quotes about the Dunkle family are from Peter Dunkle's manuscript "Thirty-Six Feet of Boys" (18 p.). Another Peter Dunkle manuscript used as background material is "Tradition and Times" (9 p.).

1. "In Memoriam—John W. Dunkle," *Pure Oil News,* March 1931; "J. W. Dunkle Dies Suddenly," *Pittsburgh Post-Gazette,* 20 January 1931; *Book of Biographies,* 160–162.

2. Page, " 'Dunkle'-Dusk-Twilight-Evening," 19.

3. *Book of Biographies,* 161.

4. Under "References," Page (" 'Dunkle'-Dusk-Twilight-Evening") writes, "George Washington Dunkle, John Wesley Dunkle, Twins, Oil Drillers, Wild Catters, Drilled first paying well in Ranger, Texas. Worked in Pennsylvania, California, New Mexico, Wyoming, Alberta." Another source is Peter Dunkle, "On Oil History." The John Wesley Dunkle whom Page identified as a twin was not Earl Dunkle's father.

5. Gille, *Encyclopedia of Pennsylvania,* 284.

6. Ibid., 490. Warren was described as a socially active city in a collection of Lester Dane Dunkle memorabilia, which includes invitations and programs. Dunkle family (DF).

7. MacGowan lecture publicity card, ca. 1906; also lecture notice, "W. L. MacGowan at Ludlow, Pennsylvania"; Warren County Historical Society, Warren, Penn.

8. Attendance and course records, Warren High School, 1900–1904, Warren County Historical Society, Warren, Penn.

9. "Our Pioneers," ca. 1905, 1 p.; lists graduates of Warren High School then attending college; Warren County Historical Society, Warren, Penn.

10. Wesley Earl Dunkle (WED), "Copper River District," 2.

11. Hammond, *Autobiography,* 1:38; Joralemon, *Adventure Beacons,* 9.

12. Davis, *Soldiers of Fortune;* Foote, *Led-Horse Claim.* Mary Foote, wife of engineer Arthur DeWint Foote, used her experiences in Colorado, Idaho, California, and Mexico to write about the life of mining men and scientists of the late nineteenth century. Mary wrote an account of her life *(Victorian Gentlewoman),* in which she responded to questions from mining engineers about her knowledge of mining men. Mary wrote (306),

> I . . . told them that I had married one of their lot and knew *them,* in their remotest hiding places, but my technique was all borrowed, and I washed my hands of any mistakes they might find. Sitting beside my own smokers in the restful silence of the Junior's room, often I thought of one of their phrases "the angle of repose" which was too good to waste on rock slides or heaps of sand. Each one of us in the Cañon was slipping and crawling and grinding along seeking what to us was that angle, but we were not any of us ready for repose.

Mary's life was fictionalized in Stegner's *Angle of Repose.*

13. Kelley, *Yale: A History,* 180–183; Chittenden, *Sheffield Scientific School,* 2:371–373. For a skeptical view of the Sheffield School's mining program in the late nineteenth century, see Spence, *Mining Engineers,* 41. On the impact of John Hays Hammond on buildings, course work, and faculty at the Sheffield School, see Chittenden, *Sheffield Scientific School,* 2:376–380.

14. The enrollment statistics were compiled from individual biographies in Church, *Class History.* On John Duer Irving, see Chittenden, *Sheffield Scientific School,* 2:381–384.

15. WED, "Class Notes."

16. Handwritten marginal notes in Le Conte, *Elements of Geology* (Yale University textbook), 243–260. WED memorabilia, DF.

17. Church, *Class History,* 375.

18. On Dunkle and his friends in the decathlon, see Howard Oliver, Letter in "1908 Class Notes." Church, *Class History,* refers to Dunkle's three-year rowing career at Sheffield (15, 20, 30, 367).

19. Baragwanath, *Good Time Was Had,* 144–145. On Baragwanath's lifestyle, his wife, Neysa McMein, and some parties, see Baragwanath, *Good Time Was Had,* 76–77, 91–93; and Sally Bedell Smith, *In All His Glory,* 109–110, 324. Details about crossing the West Fork of the Chulitna River and about Dunkle's old-age stamina are based on the Ellis interview (1996) and the Waugaman interview (1996).

20. Church, *Class History,* class biographies: Carlisle, 72; Dunham, 101; Dunkle, 102; Irwin, 152; Needham, 208; Oliver, 216; and Smith, 263.

CHAPTER 2: ALASKA IN THE GOOD YEARS

1. Koschmann and Bergendahl, *Gold-Producing Districts,* 8–9, 16–18, 26–27. For the Nome rush, see Terrence Cole, "Nome Gold Rush"; and Collier and others, *Gold*

Placers, 16–38. For the gold discovery at Fairbanks, see Terrence Cole, *Crooked Past,* esp. 35–54; and Dermot Cole, *Fairbanks,* 13–16.

2. Quinn, *Iron Rails,* 117–130, appendix 1.

3. McKay, “Copper River Area.” Excerpt from L. A. Levensaler, letter to Ralph McKay, 9 December 1955, APRD.

4. Jones, Silberling, and Newhouse, “Wrangellia.” For Dunkle’s view on the origin of the copper deposits, see WED, “Copper River District,” 6–10.

5. Dall, *Alaska and Its Resources,* 477. In regard to the Ahtna Indians, see Allen, “Expedition,” 45, 47, 49. For the Tanana Indian copper deposit, see Harris, *Schwatka’s Last Search,* 171–172, 242–243. Regarding Eskimo knowledge of native copper on Prince William Sound, see Laguna, *Chugach Prehistory,* 5.

6. Allen, “Expedition,” 22, 44–45, 49. In the summer of 1884, John Bremner traveled up the Copper River as far as the site called Taral near Chitina. On the timing and extent of the copper rush, see Grant and Higgins, “Copper Mining.” On early views of Valdez, Cordova, and Nuchek, see Lethcoe and Lethcoe, *Valdez Gold Rush,* 1, 18–20; and Janson, *Copper Spike,* esp. 1, 4–5, 10–14, 25–28.

Valdez was the entry city for the gold rush at Valdez Glacier in 1898. Cordova did not exist until about 1906, although there was a salmon cannery at Orca, about two and one-half miles from modern Cordova. Cordova itself was built approximately on the site where Eyak, another cannery, once stood. An old village, Nuchek, was west of Cordova on Hinchinbrook Island. All three locations—Valdez, “Cordova,” and Nuchek—were used by copper prospectors.

7. Slightly different accounts and dates of discovery of the Ellamar copper deposit are given in various sources. The oldest and probably best accounts are in Schrader, *Reconnaissance,* 418–420 (1900), and in Lincoln, “Big Bonanza” (1909). Schrader gives 1895 as the date when Gladhaugh discovered the Ellamar deposit and 1897 as the date when the mining claims were staked. See also Lethcoe and Lethcoe, *Prince William Sound,* 44. The Lethcoes give 1897 as the date of the Ellamar discovery and give partial credit for that discovery to Chris Pederson. Lincoln, “Big Bonanza,” 204, states that an Indian led Gladhaugh and Pederson to the outcrop of the Ellamar deposit.

For the probable discovery of the copper deposit in Horseshoe Bay, see McDonald, “Alaska Steam,” 30. Lincoln, “Big Bonanza,” 203–204, lists the locators and discovery date of the Big Bonanza or Beatson copper deposit and describes how powdered copper ore from that deposit was used as a pigment. Lincoln’s story about the stone hammers at the outcrop of the Big Bonanza on Latouche Island was corroborated by boatman Billy Pay(e) in Laguna, *Chugach Prehistory.* Laguna (5) quotes Captain Cook on the use of copper by coastal Eskimos: “[T]he natives seem to have so much of it that they refused to accept it in trade.”

8. Lincoln, “Big Bonanza,” 203–204; WED, “Copper River District,” 5; WED, “Nikolai Mine.”

9. Schrader, *Reconnaissance,* 419.

10. For recent accounts of the Valdez gold rush, see Lethcoe and Lethcoe, *Valdez Gold Rush,* 39–97; and Tower, *Icebound Empire,* 5–8. A nearly contemporaneous source of information on the rush is the account of geologist Frank C. Schrader. Schrader was attached to the Copper River Military Expedition No. 2, led by Capt. W. R. Abercrombie in 1898; his account clarifies the causes of the rush (Schrader, *Recon-*

naissance). The purpose of the Abercrombie expedition was "to find, if possible, an all-American Route, railroad or otherwise, from tidewater to the gold districts of the upper Yukon in the interior" (347). There is some confusion because most of the Valdez Glacier stampeders whom Schrader and Abercrombie met were apparently bound for the Copper River country that transportation companies and newspapers had promoted as the new but unclaimed Eldorado. The situation is almost certainly as described by Schrader (368): "[M]any prospectors and adventurers bound to the Klondike or to indefinite destinations in Alaska were led, in the season of 1898, to try their fortunes in the Copper River country. Many hoped at the same time to proceed by way of the prospective All-American route into the gold districts of the Upper Yukon."

For information on Chittyna Exploration Co. and conditions in the Copper River country during the winter of 1898–1899, see Lethcoe and Lethcoe, *Valdez Gold Rush,* 87–93; Tower, *Icebound Empire,* 70, 76–79; and Janson, *Copper Spike,* 7–10. Schrader and Spencer, *Copper River District,* 85–87, describe the location of the Nikolai copper deposit, as do DT: R2, t2 and Janson, *Copper Spike,* 8–10.

11. Tower, *Icebound Empire,* 71. Tower quotes a 1940 letter from Stephen Birch, the founder of Kennecott Copper Corp., to Alaska Territorial Governor Ernest Gruening. The letter gives a detailed account of the discovery of the Bonanza lode on 22 July 1900.

In his own account (Schrader and Spencer, *Copper River District*), Spencer mentions that he discovered the copper outcrop of the Bonanza lode about a month later than Smith and Warner, a version accepted by Douglass ("Kennecott Mines," 4). L. A. Levensaler, who entered the area only a few years after the lode was discovered, indicates that the discoveries by Smith, Warner, and Spencer were essentially simultaneous: "At the same time the property was also discovered by Mr. A. C. Spencer of the United States Geological Survey, who saw the outcrop across the Kennicott Glacier. But, by the time he arrived at the showing, it had been staked that morning by Smith and Warner" (Carlisle, "Early Days," 1). WED ("Copper River District," 22) agrees with Levensaler's version. It seems possible that, in his own account, Spencer did not want to muddy the waters with a claim of synchronous discovery. At the least, it appears that Spencer made an independent discovery of the Bonanza lode.

12. For the Wrangell Mountain discoveries and the relation of the deposits to the Nikolai and Chitistone Formations, see Mendenhall and Schrader, *Mount Wrangell District,* 28. For the discoveries around Prince William Sound, see Grant and Higgins, "Copper Mining."

13. For the origin of the Alaska Syndicate, see Graumann, *Big Business,* 7–14; Tower, *Icebound Empire,* 111, esp. 113–114; and Janson, *Copper Spike,* esp. 16–18, 121.

14. Roderick, *Crude Dreams,* 37–39. For the discovery, character, and development of the coal and oil fields near Katalla, see Martin, *Controller Bay Region,* 9, 65–66, 81–82, 112–124, including fig. 2; and Janson, *Copper Spike,* 34–37, esp. 109–117, 122–132. For the Copper River and other proposed rail routes, see Committee on the Territories, *Railroads in Alaska,* esp. map opposite p. 80.

15. The issues involved in the Cunningham case and the controversy between Pinchot and Ballinger are mainly abstracted from Graumann, *Big Business,* 11–14. For other views, see Janson, *Copper Spike,* 109–117; Evangeline Atwood, *Frontier Politics,* 209–212, 241–242; and Naske and Slotnick, *Alaska,* 83–85, 271.

16. Nichols, *Alaska,* 341.

17. Ickes, *Not Guilty,* 4–6.

18. WED, letter to Bradford Washburn, 23 March 1953, DF.

19. For a description of the "Cordova Coal Party," see Janson, *Copper Spike,* 124–127; for the dominant contemporary Alaskan view of Gifford Pinchot and his hanging in effigy, see the same reference, 123–124.

20. Evangeline Atwood, *Frontier Politics,* 209–215; Hellenthal, *Alaskan Melodrama,* 198–203.

21. For the railway battles in Keystone and Abercrombie Canyons, see Janson, *Copper Spike,* 46–48, 62–63; and Tower, *Icebound Empire,* 124–125, 129–135, 139–145. For the final selection of the route to the Wrangell Mountains, see Janson, *Copper Spike,* 26–39, 46–48, 51–54; and Tower, *Icebound Empire,* 115–116, 130–135, 146. For Mr. Barring's testimony on the difficulty of railway construction along the Copper River, see Committee on the Territories, *Railroads in Alaska,* 36–37; for the quote about Stephen Birch, see the same reference, 69.

22. Quinn, *Iron Rails,* 113–117, including photographs. For a fictional account of the railway construction project, see Beach, *Iron Trail.*

23. On Dunkle's decision to go to Alaska, see WED, letter to John Wesley Dunkle, 31 August 1908, DF; and WED, "Copper River District," 2. On the Homestead Act and the development of Alaska, see Nichols, *Alaska,* 154–159. Although the Homestead Act of 1898 (14 May) opened Alaska to homesteading, the provisions that limited homesteads to eighty acres and determined that homestead surveys were to be at private expense negated its value to most Alaskans. In Nichols's view, the most important and beneficial feature of the Homestead Act was as railway-enabling legislation. The first direct consequence was the White Pass Railroad between Skagway and the Klondike.

On the development of transportation, see Nichols, *Alaska,* 224, 231–232; and Naske and Slotnick, *Alaska,* 81–82. On Alaska home rule, see Naske and Slotnick, *Alaska,* 86–87; and esp. Nichols, *Alaska,* 383–409. The overall theme of Nichols's history of early Alaska is the battle to establish some sort of government for the new possession. The book ends with the passage of home rule legislation for Alaska in 1912.

24. Bowen, *This Fabulous Century,* 29.

25. John Whitehead, former history professor at the University of Alaska–Fairbanks and at Yale University, conversation with author, 1996. Whitehead pointed out that the annual class histories produced by Yale for many years allowed each Sheffield graduate to tell his own story to family, friends, and classmates, as in Belin, *Thirty-Year History.* Whitehead believes that there was little false modesty in the group.

26. Lord, *Good Years,* ix, 2–8; Hoover, *Memoirs,* 135–136.

27. For interesting accounts on the importance of northern gold discoveries to the national will, see Jones, "Unlikely Savior"; and Jones, "Another Spin." See also Lethcoe and Lethcoe, *Valdez Gold Rush,* 2–4. For an account of the world's greatest gold discovery, see Rosenthal, *Gold!,* author's foreword and 195–212.

28. Constructed from personal biographies and general text in Church, *Class History.*

29. Bowen, *This Fabulous Century,* 9 and tables regarding transportation.

30. Spencer, *Juneau Gold Belt,* 3, 93–97.

31. Navin, *Copper Mining and Management,* compiled from table C.

32. Joralemon, *Copper,* 233–237; and Navin, *Copper Mining and Management,* 257–260.

33. Joralemon, *Copper,* 235.

34. In Dunkle's senior year at Sheffield, he probably took more advanced courses, such as those taught by Duer Irving, which treated supergene theory more thoroughly. The earliest significant paper on enrichment (Penrose, "Superficial Alteration") was only about a decade old. Three important papers appeared in 1901 in the thirtieth volume of AIME *Transactions.* These papers expanded supergene theory and applied it to iron, gold, and silver as well as to copper deposits: Emmons, "Secondary Enrichment"; Van Hise, "Deposition of Ores"; and Weed, "Gold and Silver Veins."

35. Lindgren, "Metasomatic Processes," "Contact Deposits," and "Ore-Deposition." For an appraisal of Lindgren's views as of 1933, see Ransome, "Historical Review," 11–12; and Loughlin and Behre, "Classification of Ore Deposits," esp. 18–22, 53–54. Lindgren's views dominated hypogene theory for the first fifty years of the twentieth century.

36. On the Guggenheims, see O'Connor, *Guggenheims,* 126–135, 254–274. See also Holbrook, *Age of the Moguls,* esp. 277–302, including 295–297, which deal with the Wrangell Mountain ventures of the Guggenheims. A representative work on the oft described period La Belle Époque is Cronin's, *Paris on the Eve.* A fictional account that captures the atmosphere of the times appears in Sinclair's *World's End.*

On the diversity of gold seekers during the great California gold rush, see Borthwick, *Three (3) Years in California,* 3, 235: "The majority . . . were Americans, and were from all parts of the Union; the rest were English, French, and German. We had representatives of nearly every trade, besides farmers, engineers, lawyers, doctors, merchants, and nondescript 'young men' " (3). Borthwick mentions other groups as well: Jews, 95–96; Chinese, 41, 117–119, 215–219, 261; Missourians, 120–121, 123–124; Frenchmen, 196–198, 296–298, 300; Indians, 105–108; Negroes, 133–135; Germans, 196–198; Mexicans, 256–259, 273–277; and "Americans," 301–305. Diversity as a general characteristic of rushes worldwide was noted by Fetherling, *Gold Crusades,* 3–5; and by Morrell, *Gold Rushes,* 84, 87–88, 107–109, 122, 129, 139, 156, 205–207, 221–222, 273–274.

37. Ford, Review of "At the Actor's Boarding House"; " 'Helen Green' in Chronicle and Comment"; and Tully, "World and Mr. Nathan." On artists Laurence and Ziegler, see note 39, this chapter.

38. DT: R4, t2, sA. More on the Nowell family is given in Spencer, *Juneau Gold Belt,* 51, 58–59, 80–81; and in Roppel, "Have I Got a Deal for You."

39. Woodward, *Sydney Laurence,* 16–18, 125–126. For both Laurence and Ziegler, see Shalkop, *Eustace Ziegler,* 3–7; and Woodward, *Spirit of the North.* Shalkop (4) repeats from Carlson ("Alaska's Hall of Fame Painter") that E. Tappen Stannard purchased a mountain scene from Ziegler for $150 in 1911 (although Carlson did not specify the date). Stannard is identified as president of Alaska Steamship Co. Stannard, however, did not arrive at Kennicott until 1913 or 1914 and, when he came, would have been a new mill man (metallurgist) far down on the management chain. Stannard is also identified with Alaska Steamship Co. in 1924: "Ziegler was on a dog team with Bishop Rowe in the Chitina area when he received a telegram from E. T. Stannard offering him a commission to paint a series of murals for the Alaska Steamship Co. offices in

Seattle" (Shalkop, 4). At this time Stannard would have been vice president of Kennecott Copper Corp. in charge of Alaska mines and exploration, as well as president of Alaska Steamship Co. and the CR & NW Railroad.

40. Beach, *Iron Trail,* 368–375. In his description of the building of the Million Dollar Bridge, Beach closely follows the actual construction as outlined by engineers A. C. O'Neal and E. C. Hawkins. See Quinn, *Iron Rails,* appendix 1, 183–188.

41. Bordman, *American Operetta,* 74–75. *The Merry Widow* arrived in New York in 1907 but did not play in Paris until 1909, the same year that recordings were brought to Kennicott, Alaska.

42. The motives and effectiveness of the Alaska Syndicate have been debated since the early 1900s. A modern consensus was given by former National Park Service historian Melody Webb (Graumann):

> The evaluation of the significance of the Alaska Syndicate has been brandished about by historians for the last sixty years. It has been viewed by some as an exploitive octopus and by others as a positive economic force. Without any doubt, it played an enormous, almost overwhelming role, in Alaska's history from 1905 to 1915. The Syndicate appeared to touch every facet of life from the political to the economic, from the national to the local level. In part, Alaska's small population necessitated the impact. Its steamships carried nearly all supplies and passengers between Alaska and Seattle; its railroad was the longest and best constructed in the territory with equitable rates operating at a loss each year; its fisheries, canneries, and merchandise outlets supplied needs to a developing territory; its copper production stimulated other mineral development; and its large capital investment infused economic opportunity into an isolated area. On the other hand, the Syndicate did involve itself in politics at the local and national level that, in the aftermath given the historical perspective of the era, could be best judged as "misconduct." Overall, however, their contributions seem to outweigh their liabilities. (Graumann, *Big Business,* 23)

CHAPTER 3: APPRENTICE, SCOUT, AND ECONOMIC GEOLOGIST

Some of the material in this chapter was previously published as Hawley, "Wesley Earl Dunkle."

1. For descriptions of the education given at the Bergakadamie in Saxony, see Hammond, *Autobiography,* 1:66; and Spence, *Mining Engineers,* 30–31. On the Anaconda program, see Spence, *Mining Engineers,* 228–230; McLaughlin and Sales, "Utilization of Geology," 684; and Linforth, "Application of Geology," 695–701. Reno H. Sales continued and expanded the Anaconda program when he succeeded Horace V. Winchell as chief geologist of Anaconda in 1906.

2. Lester Dane Dunkle, postcard to WED, June 1908, DF.

3. WED, letter on six postcards to John Wesley Dunkle, 31 August 1908, DF.

4. Notes on undated postcard of Kimberley mine at Ely, Nevada, WED memorabilia, DF.

5. WED, three postcards to Mrs. John Wesley Dunkle, 4 March 1910, DF.

6. WED, postcards to family, no legible date but ca. 1909–spring 1910, DF.

7. DT: R5, t1, sA.

8. DT: R5, t1, sB.

9. DT: R6, t2, sB.

10. Lawrence ("Louis") Adams Levensaler, letter to Ralph McKay, 16 February 1967, Levensaler papers, APRD. Levensaler was with Horace Winchell in Alaska in 1903 and during the summer of 1904. He was in Butte, Montana, with Anaconda from November 1904 until February 1906. See also Carlisle, "Early Days"; Levensaler, letter to WED, 6 June 1953, APRD; and WED, letter to Levensaler, 29 June 1953, APRD.

11. Nichols, *Alaska,* 269–271; Janson, *Copper Spike,* 41, 44; Sherwood, "North Pacific Bubble."

12. Dunkle and Henry Watkins also knew Mrs. Van Campen as Helen Green Wilson (DT: R5, t2, sB), possibly from her marriage to Harry Leon Wilson, the author of *Ruggles of Red Gap* (Barry, *Seward Alaska,* 87). The marriage to Wilson is not noted in standard references on Wilson. Helen's first three marriages, in possible order, were to Bert Green (she is referred to as "Mrs. Green" in Ford, Review of "At the Actor's Boarding House"), to Wilson, and to Van Campen.

On Helen's writing career and age, see George A. Thompson Jr., New York University, letter to Dan Fleming, curator, Anchorage Municipal Library, 16 May 1991, Alaska section, Z. J. Loussac Public Library, Anchorage. See also Leonard, *Women's Who's Who,* 833. Some articles on Helen give her date of birth as 1883. According to Thompson, it is more likely that Helen was born as Helen Louise Tabour in 1880 rather than 1883, the birth date given in *Women's Who's Who.* The earlier date seems confirmed by later accounts. Helen told Fairbanks journalist Kay Kennedy in 1960 that she was eighty years old (Kennedy, "Helen Van Campen Dies"). V. Maurice Smith ("Caught in the Riffles," 17) said, "Helen Van Campen, who reads over the shoulders of newsmen and newswomen in Fairbanks with a constructive eye, reached her 78th birthday yesterday [19 February 1958]."

On Helen's writing style and her comparison with noted writers, see Tully, "World and Mr. Nathan."

13. Samuel G. Blythe, St. Petersburg, Russia, to Helen Van Campen, "on an island," 1915. Loose envelope and letter in Helen Van Campen album, APRD. On Helen as a suffragette, see Leonard, *Women's Who's Who,* 833. On Helen's unsuitability as a "Kennecott wife," refer to DT: R5, t1, sB.

14. DT: R5, t1, sB.

15. Andresen, *Memories of Latouche,* esp. 8–9.

16. DT: R5, t1, sB; see also *Cordova Daily Alaskan,* 1 December 1911.

17. WED, "Copper River District," 2.

18. Ibid.

19. Richelson, "List of Reports." Richelson lists all known reports written by staff engineers for Alaska Development and Mineral Co. between 1906 and the 1940s. From 1912 through 1915, Dunkle examined and wrote reports on more than one hundred prospects. In 1912 all the prospects described by Dunkle were in south-central Alaska. In 1913 he examined and wrote reports on thirty-five prospects in three districts in British Columbia and examined prospects in both south-central and southeastern Alaska. Richelson's records are incomplete regarding Dunkle; he lists

none of the examinations Dunkle made in 1924 that were reported to E. T. Stannard (ETS). Richelson also has no record of examinations made by Dunkle in 1928 and 1929 in British Columbia, Washington, Oregon, and Cook Inlet, Alaska, for Stannard (WED, letters to ETS, 30 October 1928 and 31 December 1928, Kennecott Exploration Co., Salt Lake City).

20. On Dunkle's first trip to New York for the company, refer to DT: R6, t2, sB. Capps (*Willow Creek District,* 52) provides an account of the Willow Creek district and notes some early discoveries. Virtually all significant discoveries in the district are described by Stoll (*Hunting for Gold,* 50, 58–61, 65–66, 74, 102). See also J. C. Ray, "Willow Creek," 204, 214.

21. W. W. Atwood, *Alaska Peninsula,* 128–129.

22. "Darkness, Lurid Light, and Thunder. Volcanic Eruption Continues Three Days," *Cordova Daily Alaskan,* 10 June 1912, 1; DT: R6, t1 continued to t2.

23. Simeon Oliver, *Son of the Smoky Sea,* 19–20.

24. On the discovery and development of the Midas mine, see Lethcoe and Lethcoe, *Prince William Sound,* 71–72; Johnson, "Jack Bay District," 168; and Johnson, *Mining,* 240.

25. DT: R4, t2, sA; DT: R5, t2, sA; and handwritten notes on photographs of 10 May 1913, DF.

26. On calculating the value of the Midas mine, refer to DT: R8, t2, sA. On the sale of the Midas mine to Granby, refer to DT: R5, t2; and Johnson, *Jack Bay District,* 168. On the transaction price, refer to DT: R1, t3. For the Midas mine, Granby paid $125,000, of which George Baldwin received $35,000 as a commission.

27. For the development of the large, low-grade, gold-in-quartz veins of the Juneau area, see Stone and Stone, *Hard Rock Gold.* For the Stokes prospects and the Granite mine, refer to DT: R8, t1, sA.

28. On the Cliff mine, Henry Watkins, George Baldwin, and the relative profitability of mines and canneries, refer to DT: R8, t1, sA and sB.

29. The approach on extralateral rights is adapted from Spence, *Mining Engineers,* 228–230.

30. Potter, "Alaska," 39.

31. WED, "Copper River District," 12–13.

32. Ibid., 10.

33. Ibid., 10–11.

34. Ibid., 18.

35. Ibid. See also Navin, *Copper Mining and Management,* 262. Navin emphasizes the complexity of the formation of Kennecott Copper Corp., which involved the assets of Guggenex, especially Braden, Chile; Bingham, Utah; and the Alaskan mines. Navin wrote,

> In operational terms the Kennecott mine was eventually as successful as anticipated, but the years required to bring it into full production dragged on so long beyond expectations that, from a payback point of view, the Alaska Syndicate was not a success. Consequently, when the price of copper rose steeply after the outbreak of World War I, the members of the Syndicate decided that an opportune time had come to unload their investment. Believing that the public would pay a higher price for a diversified property than for one located only in the mountains of Alaska,

the Guggenheims agreed to pool the assets of Guggenex as well. This meant forming a company that would have a major mine in Chile; . . . a minority interest (37 percent) in the second largest copper mine in the United States (Bingham Canyon did not outstrip the mines at Butte until the late 1920s); and a rich but small mine in Alaska. The company was called Kennecott, and at first its entire stock issue was exchanged for the holdings of the members of the Alaska Syndicate.

36. I first heard a version of this story from Hugh E. Matheson, a mining man from Denver, Colorado, who grew up and mined in Alaska early in his career. In turn, Matheson heard the story from his father, who knew Dunkle from the 1920s. In Matheson's version, the punch line was "I didn't know we had one of those [copper mines] up there."

37. Bateman and McLaughlin, "Geology of the Ore Deposits," 4.

38. Ibid, 54–56.

39. Kennecott Copper Corp., "Profile of Jumbo Tram," surveyed by Graeff-Dunkle-Hancock, drawn by C. P. Hazelet, 7 July 1914. Engineering drawings, McCarthy Quadrangle, number 03, APRD. On the Dunkle and Hinckley studies on leaching of the copper carbonate ores, refer to DT: R5, t2, sB; on the ventilation problem, refer to DT: R8, t2, sB.

40. Bleakley, *Chisana Mining District,* 11–12.

41. On Dunkle's arrival in and study of the new district, see WED, "Chisana Report, 9 December 1914," Kennecott Exploration Co., Salt Lake City. On the McClellan incident, refer to DT: R5, t1, sA.

42. DT: R5, t2, sA.

43. DT: R5, t1; WED, "Copper River District," 19.

44. DT: R1, t2.

45. *Men of Sandusky,* 1895, 50; *Bench and Bar of Ohio,* 1897, 379–380; *History of Erie County,* 1889, 584–585 (incomplete references furnished by Sandusky Library, Sandusky, Ohio). On the death of Hull, see "In Memoriam: Erie County Bar Pays the Last Tribute of Respect to the Late Judge Hull," *Sandusky Star Journal,* 5 June 1905; "Circuit Judge Linn W. Hull Dies at Battle Creek, Michigan," *Sandusky Daily Register,* 28 May 1905, 1; "Last Mark of Respect," *Sandusky Star Journal,* 29 May 1905; "Judge Linn Walker Hull," *Sandusky Daily Register,* 25 July 1905, 3; and H. L. Peeke, Esq., "Bar Honors Dead Judge," *Sandusky Daily Register,* 6 June 1905, 1, 3.

46. Brandt, *Town That Started the Civil War,* esp. 25–49. See also Fletcher, *Oberlin College,* 1:142–149, 170–178; 2:516–517, 523–536, 771–773.

47. Peeke, "Bar Honors Dead Judge," 1, 3.

48. Oberlin College, "Florence Hull," unnumbered photograph and legend.

49. Oberlin College, Office of the Secretary, 30 April 1931 (received 21 May 1931), death records for Florence Hull prepared by Oberlin College; see also Mrs. Marguerite Hull Badger, note for alumni necrology; Oberlin College archives.

50. "Culmination of Alaska Romance," *Oberlin Review,* 25 December 1914.

CHAPTER 4: LIFE AND GEOLOGIC CHALLENGES AT KENNICOTT

1. A more analytic view of Kennicott and its social stratification appears in Cronon, "Kennecott Journey," 44–50.

2. DT: R1, t2.

3. Graumann, *Big Business,* 27, 31, 35–36.

4. DT: R6, t2, sB.

5. One celebration of the Fourth of July was described by Henry Watkins (DT: R6, t2). The Chisna Development party reached the Tonsina Military Bridge on 4 July 1900 and celebrated with a musical recreation of the original Fourth of July in 1776. On Seagrave's schedule change, refer to DT: R1, t3.

6. On socializing among the Kennicott mine staff during the early days, refer to DT: R4, t2, sA and sB. On later social events for the entire community, see Ricci, "Childhood Memories of Kennecott."

7. On Larry Bitner and the arrival of the first women at Kennicott, refer to DT: R1, t2; and DT: R4, t2, sA and sB.

8. Shedwick, Untitled typed journal. The journal tells of Larry Bitner's marriage (note 7), company staff, Shedwick's courtship, and the hazing of new employees and newlyweds at Kennicott (120, 126, 130, 140).

9. DT: R4, t2, sB.

10. On the annual baseball game, see Berg, "Rhinehard Berg," pt. 1, no. 3, p. 11. On Carl Engstrom and his artistic endeavors with the copper ore, see L. A. Levensaler, letter to Ralph McKay, 26 August 1965, Levensaler papers, APRD.

11. DT: R4, t2, sA; 1920s address book, WED memorabilia, DF.

12. On Dunkle's hunting interests and Stephen Birch's sheep hunt at the start of World War I, refer to DT: R4, t2, sB, and DT: R5, t1, sB.

13. On the value of Shedwick's Kennecott stock, see Shedwick, Untitled typed journal, 181–182. On Dunkle's Kennecott share interest, refer to DT: R5, t2, sA; and work sheets for income tax return, 1937, WED memorabilia, DF.

14. Douglass, "Kennecott Mines," 7.

15. On the engineer's tram rides, refer to DT: R1, t1.

16. DT: R1, t3; DT: R5, t1; L. A. Levensaler, letters to Ralph McKay, esp. 1 July 1965, 26 August 1965, and 2 July 1968, Levensaler papers, APRD.

17. DT: R6, t2, sB.

18. Joralemon, *Copper,* 57–67, 225–227.

19. On comparative per-man copper production from the Kennecott mines and Butte, see Douglass, "Kennecott Mines," 6–7. On the production of electrical power from Kennicott tramlines, see Douglass, "Kennecott Mines," 6–7; and DT: R1, t1. On the initial high-grade production of ore from the Kennecott mines, see Joralemon, *Copper,* 226–227: "As soon as the railroad was finished the millions came rushing back. For a time Kennecott became the bugaboo of the copper producers. As production kept mounting, the rival companies feared that the Bonanza ore would wreck the copper market of the world. New ore bodies almost as rich as the first one added to the fear. Then fate relented. Ore reserves dwindled and grade of ore fell."

20. Graton and Murdoch, "Sulphide Ores of Copper"; Bateman and McLaughlin, "Geology of the Ore Deposits," 76–77.

21. On Dunkle's theory of the origin of the Kennicott ore, see WED, "Copper River District," 6–10. On the industrial espionage carried out by Anaconda Co. and an Anaconda expedition to Alaska, see data from the cover pages of Julia Sweeney, "Reconnaissance," album 3, Levensaler papers, APRD; and Carlisle, "Early Days." On Horace Winchell's view on the origin of chalcocite, see Carlisle, "Early Days."

22. W. H. Emmons, "Supergene Enrichment," 405–406.

23. Zies, Allen, and Merwin, "Secondary Copper Sulphide Enrichment"; Posnjak, Allen, and Merwin, "Secondary Enrichment."

24. Bateman and McLaughlin, "Geology of the Ore Deposits," 35–40, plate 3.

25. Craig, "The Cu-S System," CS58–CS76. The mineral djurleite is chemically very close to chalcocite. Another mineral of near-chalcocite composition, digenite, is a solid solution of chalcocite and covellite that is possibly stabilized by a very small amount of iron. Chalcocite itself forms four temperature phases, none of which coincide with those proposed by the earlier investigators. For comments on the mineralogy of ore, see also MacKevett and others, "Kennecott-Type Deposits," 79–80.

26. DT: R1, t1, and DT: R2, t2. See also WED, "Copper River District," 24.

27. WED, "Copper River District," 24; Carlisle, "Early Days"; WED, letter to Stephen Birch, 26 December 1913, Packet C, File 140, Document 7, Kennecott Copper Corp., Salt Lake City.

28. WED, "Copper River District," 24; WED, letter to Mr. and Mrs. John Hull Dunkle, 11 February 1957, DF.

29. WED, "Copper River District," 15–16.

30. Ibid., 17.

31. Keller, "Mother Lode"; Carl Ulrich, letter to Stephen Birch, 24 July 1912, Packet C, File 140, Document 6, Kennecott Copper Corp. files, Salt Lake City.

32. WED, letter to Stephen Birch, 26 December 1913, filed as above, Document 7.

33. Carl Ulrich for Stephen Birch, letter to WED, 6 January 1914, filed as above, Document 8.

34. Stephen Birch, letter to WED, 6 January 1914, marked "Confidential," filed as above, Document 9. Kennecott files indicate that this letter is from Carl Ulrich. At this time, Ulrich was Birch's young secretary; he had never been to Alaska, did not know Seagrave, and could not give orders. The letter was from Birch, who dictated it to Ulrich.

35. Simpson, "Bonanza-Motherlode Mine," 8.

CHAPTER 5: SCOUTING FOR BIRCH

1. Richelson, "List of Reports." Statistics on properties examined by Dunkle were constructed from Richelson's report.

2. Information on early placer discoveries in Alaska can be found in several sources. On the Hope-Sunrise district, see Buzzell, *Memories of Old Sunrise,* xi–xiv. On the Willow Creek district, see Stoll, *Hunting for Gold,* 24–25; Paige and Knopf, "Matanuska and Talkeetna Basins," 104–125; and Capps, "Western Talkeetna Mountains," 188, 199. On the Yentna or Cache Creek district, see Sheldon, *Heritage of Talkeetna,* 8–12. On the Chistochina (Chisna) district, see Tower, *Icebound Empire,* 48, 182; and Lethcoe and Lethcoe, *Valdez Gold Rush,* 112, and 9–10 for an account of Capt. West. West probably discovered gold in the Chistochina country in the 1880s. On the discovery of the Nizina district, refer to DT: R4, t1.

3. On the Wells brothers, see "Death Takes A. O. Wells," *ADT,* 21 February 1945, 1; and "Final Tribute Paid Mr. Wells," *ADT,* 26 February 1945, 3. On the Wells brothers' partner John Coffey and the location of the Golden Zone claims, see the Talkeetna Recording District claim records, Book 1, Palmer, Alaska; and Capps, "Upper Chulitna Region," 208. On the claim activity in the Broad Pass region up to November 1914, see "Who's Who in the Broad Pass Region," *Knik News,* 27 March 1915, 1.

In regard to Dunkle's assignment to visit the new district and his meeting with Tom Mack, refer to DT: R8, t1, sA. On lease transactions by Thomas Aitken, see "Who's Who in the Broad Pass Region." The *Knik News* spelled the name as "Aiken," but it is almost certainly Thomas P. Aitken, a Scottish immigrant who had notable success in Cripple Creek, Colorado, and in Fairbanks and Flat, Alaska, and who often dealt with the Guggenheims.

4. DT: R8, t1, sA.

5. Ibid.

6. On the return of Earl and Florence Dunkle to Alaska, see "Mining Engineer Returns With Bride," *Cordova Daily Alaskan,* 29 January 1915. See also Clifton, *Rails North,* 69–71, 76–78, for accounts of the Alaska Northern Railroad, which at that time ended at Kern.

7. DT: R6, t2, sB.

8. Ibid.

9. Ibid.

10. Ibid.

11. Item in short note: "W. E. Dunkle, mining engineer, departed Tuesday for the Broad Pass Country. Jack Cronin, with his dog team, is taking him" (*Knik News,* 27 March 1915, 1); see also "Broad Pass Is Booming," *Knik News,* 6 March 1915, 1. In regard to the chance meeting of Dunkle and the Nugget Kid, see WED, letter to Bruce Borthwick Dunkle (BBD), 20 June 1955, 1, DF. On the commercial dog mushers used in the operation, see "Many Outfits in the Broad Pass," *Knik News,* 10 April 1915, 1.

12. "Broad Pass Making Good," *Cook Inlet Pioneer,* 26 June 1915.

13. "Preliminary Work on Great Railway Project Begins," *Knik News,* 1 May 1915, 1.

14. "Sourdoughs Assemble in the Coal Terminal," *Cook Inlet Pioneer,* 12 June 1915, 3. See also Crittenden, *Get Mears!,* esp. 80–99.

15. See note 12, this chapter.

16. DT: R3, t2.

17. Capps, "Upper Chulitna Region," 208–210, 216–217.

18. Browne, *Conquest of Mt. McKinley,* 221. One of the Wells brothers had already recognized the need for better transportation to the area, as he told Browne: "If you tear off anything for the papers about this neck o' the woods tell 'em we need a railroad and we need it bad" (221). See also Dunkle's notes on back of Wells's photograph, WED memorabilia, DF.

19. On Dunkle's schedule, see "Brevities," *Cook Inlet Pioneer,* 24 July 1915, 4: "W. E. Dunkle . . . was in town this week." On the remarks of William Springer, see "Thinks Well of Broad Pass Country," *Cook Inlet Pioneer,* 11 September 1915. Based on the author's samples collected in the Guggenheim tunnel at the Golden Zone mine, the average concentration in the main drive was about 0.08 ounces of gold per ton; the average in a side drift was about 0.07 ounces of gold per ton. On the necessity of railroad access, see Moerlein, "Golden Zone Property Evaluation."

20. On fluctuating copper prices during World War I, see Navin, *Copper Mining and Management,* 121, 123, appendix, table C. On Stannard's arguments for a management change at Kennicott to augment copper production and its implications on the mine staff, refer to DT: R5, t1 and t2.

21. DT: R5, t2; Douglass, "Kennecott Mines," 10.

22. DT: R5, t2, sA; Graumann, *Big Business,* 25.

23. DT: R5, t2, sA.

24. Ibid. See also Douglass, "Kennecott Mines," 10. For a general review of the history of flotation, see Bunyak, "To Float or to Sink," 35–44.

25. DT: R5, t2, sA; Douglass, "Kennecott Mines," 10.

26. Duggan, "Ammonia Leaching at Kennecott"; Spude and Faulkner, *Cordova to Kennecott, Alaska,* 40–41.

27. Bateman and McLaughlin, "Geology of the Ore Deposits," 45, 49. On total mill recovery, see Douglass, "Kennecott Mines," 10.

28. DT: R5, t2, sA.

29. On the position of Stadtmiller at Beatson, refer to DT: R5, t1, t2, and t3. On the later employment of DeWitt Smith and David Irwin, see "H. DeWitt Smith," in *Who's Who in Engineering* (1940), 1650; "David D. Irwin," in *Who's Who in Engineering* (1941), 897; and Carlisle, "David D. Irwin," 88–91.

30. Regarding Stannard's personality, see Graumann, *Big Business,* 29–30, 51. On the offer to strike made by Dunkle's miners, refer to DT: R5, t2, sA.

31. William E. Dunkle, letter to author, 11 March 1997.

32. On Stannard and the termination of Stadtmiller, respectively, see Graumann, *Big Business,* 29, and DT: R5, t2, sA.

33. For the history and production of the Contact district, see LaPointe, Tingley, and Jones, *Mineral Resources of Elko County, Nevada,* 61–66; and Schrader, "Contact Mining District," 22–26. For the addresses of Dunkle's Contact associates, see Dunkle's 1920s address book, WED memorabilia, DF.

34. Internal memorandum, Hirst-Chichagof Mining Co., 9 December 1925, Hirst-Chichagof files, APRD. The memorandum mentions Dunkle's salary of one thousand dollars per month, which must have been very high for the times.

CHAPTER 6: FROM PIER 2, PORT OF SEATTLE, TO AFRICA

1. ETS, letter to Stephen Birch, 19 May 1920, Kennecott Exploration Co., Salt Late City. Stannard noted that Dunkle had arrived in Seattle on 10 May, had obtained a house, and was ready to work.

2. Spence, *Northern Gold Fleet,* 57.

3. DT: R4, t1; DT: R5, t2.

4. DT: R4, t1; WED, telegram to ETS, 5 August 1920, filed as note 1 above.

5. Report on Anderson property, WED, letter to ETS, 3 August 1920, 4, originally from Kennecott Exploration Co., Spokane, Wash.

6. On Dunkle's alternatives on the return trip from Candle to Nome, refer to DT: R4, t1; WED, telegrams to ETS, 4 and 5 August 1920, filed as note 1 above; and "Passengers of Lost Schooner *White Mountain* Arrive," *Nome Nugget,* 7 August 1920, 1–2.

7. On the results of the Anderson property evaluation and the shipwreck near Lopp Lagoon, see WED, telegrams to ETS, 4 and 5 August 1920, filed as note 1 above.

8. ETS, letter to WED, 4 August 1920, filed as note 1 above.

9. WED, letter to ETS, 13 August 1920, filed as note 1 above.

10. DT: R3, t2, sA.

11. DT: R7, t1; Tower, *Icebound Empire,* 255; *Cordova Daily Times,* 21 February 1923.

12. The date of Dunkle's recommendation of the Lucky Shot to Kennecott or possibly ASARCO is uncertain. In the tapes, Dunkle says 1922, but he also states that he recommended Lucky Shot to L. C. Thomson after the property was turned down by Kennecott. Thomson took over at Lucky Shot about 1920, so the date of recommendation to Kennecott may have been earlier, perhaps 1920.

Dunkle invested in Lucky Shot stock. He bought one thousand shares in October 1922 and another 178 shares sometime before the end of 1924. The investment was consistently profitable. In 1926 he made $82.50 on his investment of $612.26; in 1927 his profit was $424.08 ("L" listings, 1920s address book, WED memorabilia, DF).

13. The Mabel was optioned by Alaska Development and Mineral Co. for a total purchase price of $75,000 in 1922. An initial cash payment of $4,000 was made on 18 July 1922. Mabel Co. (Mabel Mining, Milling, and Power Co.) was capitalized at $100,000, and $88,000 of the stock was subscribed. Caveny (called "Bill" by Dunkle and "Tom" by William Stoll, *Hunting for Gold*) was the major shareholder at $30,000. The Alaska Development lease also called for payments of $11,000 and $15,000, respectively, on 1 October 1924 and 1 October 1925. The balance was due in 1926. Part of the payments could be made from the proceeds of the mine ("M" listings, 1920s address book, WED memorabilia, DF; Stoll, *Hunting for Gold,* 74). See also "Engineer Arrives in Anchorage to Start Exploring," *FDNM,* 23 February 1923, 1.

14. WED, letter to Bradford Washburn, Boston, 20 March 1953, DF.

15. Ibid.

16. Ibid.

17. Stoll, *Hunting for Gold,* 74–80; WED, letter report to ETS on mining properties visited in 1924, December 1924, filed as note 1 above.

18. William E. Dunkle and John Hull Dunkle, Dunkle family interviews, 1997.

19. WED, letter report to ETS on mining properties visited in 1924, December 1924, filed as note 1 above.

20. Internal memorandum, Hirst-Chichagof Mining Co., 9 December 1925, Hirst-Chichagof files, APRD. The memorandum mentions that Rust of the Chichagoff mine near Sitka, Alaska, was interested in the Nevada-Bellevue mine at Contact, Nevada. The investment in the purchase and redevelopment of the mine is described in Schrader, ("Contact Mining District," 23.

21. This section is based on Schrader, "Contact Mining District"; LaPointe, Tingley, and Jones, *Mineral Resources of Elko County;* and as cited in notes 23 and 24 below, Kennecott correspondence. Production summaries are from LaPointe, Tingley, and Jones; most of the detail for the 1925–1930 operation of the Nevada-Bellevue mine is from Schrader (22–26). The Dunkle sons remembered the acid stream in front of their house.

22. William E. Dunkle, letters to author, 11 March 1997 and 4 May 1997; William E. Dunkle, interview, 1997.

23. WED (in Contact, Nev.), letter to ETS (in Seattle), 16 May 1927, filed as note 1 above.

24. This paragraph is based on a series of telegrams and letters between ETS, Stephen Birch, and WED, including ETS, coded telegram with translation to WED, 21 May 1927; ETS, letter to Stephen Birch (New York), 21 May 1927; Stephen Birch, telegram to ETS, 25 May 1927; and Stephen Birch, telegram to ETS, 31 May 1927; filed as note 1 above.

25. WED, letter to ETS, 1 September 1927, filed as note 1 above; DT: R8, t1, sA.

26. Robert H. Redding (son of Emily [née Hull] Redding), letter to author, 3 October 1997; Florence and Marguerite Hull files, Oberlin College records, Oberlin, Ohio; Redding, *North to the Wilderness,* 12–14.

27. Cover of "M" listings, 1920s address book, WED memorabilia, DF.

28. WED (in Contact, Nev.), letter to ETS (in Seattle), 16 October 1926; WED (in Contact), letter to ETS, 31 October 1926; WED (in Contact), letter to ETS, 14 February 1927; WED (in Ketchikan), letter to ETS, 18 June 1928. Related letters include WED (in Contact), to W. A. Dickey, 31 October 1926; WED, to Stephen Birch, 11 July 1928; and WED, to Stephen Birch, 7 November 1928. See also ETS, memorandum to Stephen Birch, 16 November 1928. All filed as note 1 above.

29. Navin, *Copper Mining and Management,* 120.

30. Ibid., 343–345; Joralemon, *Copper,* 272–273, 278–279, 281, 283. The first European prospectors found hundreds of ancient mining sites in the copper belt of southern Africa. One site was a pit seven hundred feet long, four hundred feet wide, and thirty feet deep, from which African miners had extracted a half million tons of oxidized copper ore. The Star of the Congo mine was opened on one of the ancient African copper pits.

31. Navin, *Copper Mining and Management,* 353–354; Joralemon, *Copper,* 285, 287.

32. Navin, *Copper Mining and Management,* 276, 288.

33. WED, postcards to John Wesley Dunkle, Mrs. John Wesley Dunkle, and Lester Dane Dunkle, 27 February 1929; WED, postcard to John Wesley Dunkle, 4 March 1929; DF.

34. WED, postcard to John Wesley Dunkle, 4 April 1929, DF. Dunkle wrote, "Am on my way to Jo'burg this AM."

35. WED, "Class Notes."

36. Lindgren, *Mineral Deposits,* 241.

37. Dixon, "Platinum Deposits."

38. Dixon, "Luanshya Copper Deposit." Dunkle was at Elisabethville in early May. He sent several postcards from there to family members in Pittsburgh on 5 May 1929.

39. Carlisle, "David D. Irwin"; "David D. Irwin," in *Who's Who in Engineering* (1941), 897.

40. Navin, *Copper Mining and Management,* 353–357, table B, 396–397.

41. DT: R5, t3.

42. DT: R5, t1, sA; DT: R5, t3; "H. DeWitt Smith," in *Who's Who in Engineering* (1940), 1650. For conditions in Africa at that time, see Navin, *Copper Mining and Management,* 278–279, 289–290, 292. Earlier spellings for the two Newmont properties are O'okiep and Palabora.

43. William E. Dunkle, letter to author, 15 December 1995. Earl Dunkle returned to Elisabethville in September 1929. He sent a series of postcards from there to Florence and his sons in Seattle and to his parents and brother in Pittsburgh on 2 September 1929 (DF).

44. DT: R5, t3, sB.

45. Navin, *Copper Mining and Management,* 135, 266–267.

46. DT: R5, t2 and t3.

47. Gladys ("Billie") Borthwick Rimer (later Dunkle) graduated from St. Andrews University in 1924 (Norman H. Reid, St. Andrews University, letter to author, 14 May 1996). She completed her Ph.D. in 1931 at the University of Cape Town. Her thesis

was on the endocrine glands of a South African toad (Kathleen Clark, Alumni Office, University of Cape Town, letter to author, 11 July 1996). Billie's major professor was Lancelot Hogben, a world authority on the ductless glands.

48. DT: R5, t3, sA.

CHAPTER 7: PARDNERS MINES AND THE LUCKY SHOT

1. The material on Harold E. Talbott was compiled from several sources. The basic outline is from "Harold E. Talbott," in *Who Was Who in America,* v. 4 (1942). Most of the material about Talbott before 1916 is from Thayer and others, *Katherine Houk Talbott,* esp. 167–168, 226–227. Harold Talbott's involvement with Pardners Mines is mostly from Baragwanath, *Good Time Was Had,* 126–127. Material about Talbott's later life is from Zimmerman, "Commencement Weekend," cover photograph and 3, 5. A citation that Harold Talbott received for an honorary LL.D. says, "A former associate of Orville Wright and producer of many of the war planes used by our forces in World War I, he is one of those whose courage and imagination have made possible the conquest of the skies, in peace and war." The citation also notes Talbott's long service to the Republican Party (Archives, Hobart and William Smith Colleges). In 1955 Talbott was secretary of the Air Force.

2. "Philip Danforth Wilson," in *Who's Who in Engineering* (1940), 1958–1959.

3. Baragwanath, *Good Time Was Had,* 62.

4. Ibid., 75–76, 83. On the life style and attainments of Neysa McMein, see Sally Bedell Smith, *In All His Glory,* 109; and Marie Beynon Ray, "Throwing a Party." See also Coward, *Future Indefinite,* 126; and Coward, *Present Indicative,* 176. On Bernard Baruch as financial advisor to McMein, see Baragwanath, *Good Time Was Had,* 106.

5. On his dabbling in art, see Baragwanath's *Good Time Was Had* (159–161) and *Pay Streak.* On reviews of Baragwanath's memoirs and play, see "Mining Engineer"; Baragwanath, *Good Time Was Had,* 179–182; Union Catalog Pre-1956 Imprints; and Review of *All That Glitters.*

6. There were two nude paintings in the Mining Club. One, a Neysa McMein, hung over the bar; the other, by Baragwanath, hung in the men's room. The first "was art, the second, sex." The Mining Club has now combined with the Chemists' Club; the pictures are in storage in a "barn in Long Island." The information and quotes above are from George Kruger, mining engineer, Pleasantville, N.Y., telephone conversation with author, spring 1996. On Baragwanath and Bill Paley's enjoyment of Baragwanath's stories, see Sally Bedell Smith, *In All His Glory,* 324, 329–330.

7. Baragwanath, *Good Time Was Had,* 127; Bennett, *Quest for Ore,* 278–280. Bennett's account does not confirm Baragwanath's report on the Bulolo dredging project but is compatible, as indicated in the quote below. About half the money needed for the project was raised in Australia. Bennett noted,

> [Charles A.] Banks then went to the United States where he, with Frank Short, attempted to raise the remainder. The times were not propitious. It was early in 1930 and the shadow of the Great Depression was lengthening over the land. One would have thought that the financial community in New York would have been quick to see the interrelationship between falling wages and prices and a fixed value for gold. . . . [B]y persistence the sum needed was finally subscribed. (278–279)

Harold Talbott's Pardners Mines presumably contributed to the subscription.

8. DT: R5, t3, sB.

9. Baragwanath, *Good Time Was Had,* 142–143.

10. DT: R5, t3, sB.

11. William E. Dunkle, telephone conversation with author, 1997.

12. Baragwanath, *Good Time Was Had,* 143.

13. Koschmann and Bergendahl, *Gold-Producing Districts,* 12–13; Stoll, *Hunting for Gold,* 284.

14. Stoll, *Hunting for Gold,* 26.

15. On the discoveries of gold at the Independence, Gold Bullion, and Gold Cord, see Stoll, *Hunting for Gold,* 50, 74, 92. On the discovery of gold at the Lucky Shot and War Baby, see Stoll, *Hunting for Gold,* 102.

16. On the Lucky Shot and L. C. Thomson, refer to DT: R5, t3, sA. On production from the Gold Bullion, see Stoll, *Hunting for Gold,* 56–59.

17. Stoll, *Hunting for Gold,* 103–104.

18. Ibid., 105.

19. DT: R5, t3, sA and sB. On examinations of the Lucky Shot in 1929 by Connell and McAllen, see "Development of Coast Property Predicted," *FDNM,* 17 May 1930, 2.

20. Item in "Alaska Notes," *FDNM,* 19 March 1930, 2.

21. "Development of Coast Property Predicted," *FDNM,* 17 May 1930, 2.

22. On development of the Lucky Shot in 1930, see P. S. Smith, "Mineral Industry of Alaska," 16–17.

23. William E. Dunkle, to author, Dunkle family interviews, 1997.

24. Stoll, *Hunting for Gold,* 109.

25. Ibid., 106–107; J. C. Ray, "Willow Creek," 206–207.

26. Stoll, *Hunting for Gold,* 108.

27. Joralemon, *Adventure Beacons,* 310–312.

28. "Death of Florence Hull Dunkle," *ADT,* 14 May 1931.

29. William E. Dunkle, letter to author, 11 March 1997; Dunkle family interviews, 1997.

30. WED, postcard to Mrs. John Wesley Dunkle, 15 May 1931, DF. This postcard was written in Chicago, almost certainly after a brief visit to Pittsburgh following Dane Dunkle's death on 16 April 1931. WED wrote, "Spent today here writing letters, 19 of them, and two telegrams, so have been busy. On my way tonight."

31. The material on gold prices and related events is generally constructed from Morris, *Encyclopedia of American History,* 338–339, 341–342, 344–346; Stoll, *Hunting for Gold,* 116–118; and DT: R4, t1, sB. The events are also discussed in Bernstein, *Power of Gold,* 320–323.

32. DT: R4, t1, sB.

33. Schwarz, *Speculator,* 270–279, 295–298.

34. Baragwanath, *Good Time Was Had,* 144–145; DT: R9, t1. In Dunkle's account of the trip, Neminen would not even make a low pass at the Lucky Shot to see what the landing conditions were like. Dunkle did not think too much of Neminen as a pilot.

35. On gold production from the Lucky Shot and plans for continued development of the mine, see Stoll, *Hunting for Gold,* 111–113.

36. Baragwanath, *Good Time Was Had,* 184–188, 197–199, 207–208, 213, 219.

37. Stoll, *Hunting for Gold,* 113, 115.

38. Luther Noey, interview, 1996. On Dunkle's income from the Lucky Shot and his tax calculations for the year 1937, see Dunkle's tax return, WED memorabilia, DF.

CHAPTER 8: ALASKA'S FLYING MINER

1. On initial flights in the Fairbanks area, see Stevens, *Alaskan Aviation History,* 1:123–126; on flights near Nome, see 137–142 in the same volume. On mining-related flights out of Anchorage and on Russel Merrill, see MacLean and Rossiter, *Flying Cold,* 163–170; and Stevens, *Alaskan Aviation History,* 2:646.

2. DT: R9, t1, sA.

3. On Dunkle's initial flight training and the founding of Star Air Service, see Satterfield, *Alaska Airlines Story,* 11–15; and DT: R9, t1, sB. On the naming of the air service, see Satterfield, *Alaska Airlines Story,* 12.

4. On Dunkle's completion of flight training and licensing, see Mills and Phillips, *Sourdough Sky,* 165; and Dunkle's pilot's license, WED memorabilia, DF.

5. The account of Dunkle's first transcontinental flight is mostly on DT: R9, t2, sA, with some material from sB and from DT: R9, t3, sA.

6. Baragwanath, *Good Time Was Had,* 146.

7. Brown, *Central City,* 121.

8. DT: R9, t2, sB; Porter, *Flying North,* 206–207.

9. DT: R9, t2, sB.

10. Ibid.

11. On Dunkle's use of Travel Air and on its final crash, refer to DT: R9, t2, sB; and "Dunkle Not Hurt in Plane Crash," *FDNM,* 13 June 1936, 3. On the utility of the Aeroncas and the sale of one to Bill Egan, refer to DT: R9, t2, sB. On the utility of airplanes in prospect examinations, see "Mine Operator on a Solo Hop to Kobuk," *FDNM,* 8 July 1933; and "W. E. Dunkle Returns From Kobuk," *FDNM,* 10 July 1933, 8.

12. The account of Dunkle's second transcontinental trip is mostly from DT: R9, t2, sB.

13. Material on Dunkle's flights near Washington, D.C. is derived from DT: R9, t2, sB; "Dunkle Tours States in His Own Airplane," *FDNM,* 29 March 1934, 4; and Bartlett, "Alaska's Air Needs," 3.

14. DT: R9, t2, sB.

15. Ibid. See also "Dunkle Makes Ready to Fly Coast Route," *FDNM,* 13 April 1934, 6.

16. "Record Hop Made Seattle-Anchorage by W. E. Dunkle," *FDNM,* 8 May 1934, 5; DT: R9, t3, sA.

17. William E. Dunkle, *Birds Eye View,* 72–77.

18. Mills and Phillips, *Sourdough Sky,* 137–138.

19. There are several reasons to believe that Dunkle went to London in 1934 to court Billie Rimer. Jack Neubauer, who was L. McGee's (McGee Air Service) mining partner for many years, thought that Dunkle and Billie were married in London. But the Dunkles were married in Valdez, Alaska, in August 1935. Harry Buhro, cited in note 20 below, thought that Dunkle traveled to South Africa to meet Billie in 1934, but Billie was in the British Isles after 1933. Diana Rimer Edwardes writes that she and her father left London for South Africa in the summer of 1934 (letter to author, 24 September 1998) and that Billie, who was supposed to follow them, never returned to Africa. By combining the stories, one can assume that Earl met Billie again in the

summer of 1934 in London and brought her to Alaska after that. Billie never saw her daughter again, but mother and daughter corresponded, and Billie was very proud of her daughter's accomplishments. The older Dunkle sons, Jack and Bill, had little contact with their father for several years after Florence's death and could add little to the picture. Bill found out about his father's second marriage when an instructor at Menlo School in California brought in a clipping (Bill thought it was from the *New York Times*), which stated that his father had married a Mrs. Rimer.

20. Harry Buhro was born in Russia in about 1875. His own life was probably the basis of his novel, a minor Alaskan and sailing classic: *Rough-Stuff and Moonlight.* See also Harry Buhro, letter to Mrs. Dunkle, 29 September 1953, DF. Buhro and Billie corresponded for many years. Billie grubstaked Buhro in his search for gold and diamonds in Alaska during the late 1930s.

21. "W. E. Dunkle Flies Waco Here With 2 Passengers," *FDNM,* 19 September 1934.

22. Mrs. Frank (Maxine) Reed Jr., conversations with author, 1997 and 1999. See also Agnes (no family name), Pittsburgh, letter to WED, 6 October 1937, DF.

23. DT: R9, t1, sA.

24. On Dunkle's flying skill and his flights with William Dunkle, see William E. Dunkle, letter to author, 22 April 1998. On Dunkle's freedom from airsickness and vertigo, refer to DT: R6, t2, sA; and author's interviews of William E. Dunkle and Bruce Borthwick Dunkle.

25. WED, letter to Gladys ("Billie") Rimer Dunkle (GBD), 19 August 1938, DF.

26. On Dunkle's flights through Anderson Pass and an emergency landing, see Porter, *Flying North,* 206–207. On landing at Bill Shannon's prospect, see Bill Shannon, letter to WED, 26 May 1936, DF; and WED, letter to John Baragwanath, 10 September 1936, DF. On Dunkle's failed attempt on takeoff with Al Hamberg, see WED, letter to Bradford Washburn, 20 March 1953, DF; on his uphill takeoff from Shannon's field, see Porter, *Flying North,* 206–207.

27. WED, letter to Bradford Washburn, 20 March 1953, 2–3, DF.

28. Ibid., 7.

29. DT: R3, t1.

30. "Plane Takes Tractor Over Alaska Range," *FDNM,* 22 April 1937.

31. On the relationship between Dunkle, Star Air Service, and the Lucky Shot mine, see Satterfield, *Alaska Airlines Story,* 18, 31. Information on the flying support that Star Air Service provided for other Dunkle operations comes from Ledger A, "Investment Accounts," DF. (See the initial general note for chapter 9, "A Sure Cure for Depression.")

32. Stevens, *Alaskan Aviation History,* 1:123; Satterfield, *Alaska Airlines Story,* 3–4; "L. Mac McGee," obituary, *Anchorage Daily News,* 18 June 1988; Jack Neubauer, interview, 1996; DT: R9, t1, sB.

33. Satterfield, *Alaska Airlines Story,* 31; DT: R9, t1, sA; William E. Dunkle, letter to author, 22 April 1998. Bill Dunkle believed that Steve Mills, then Kenny Neese, managed Star Air Service, but in the tapes WED confirms that McGee temporarily managed Star after it bought McGee Airlines. Because of Dunkle's dominant share in Star Air Service, it seems likely that much of the fifty thousand dollars to buy out McGee ultimately came from the Lucky Shot mine, but through Dunkle's pocket (DT: R9, t1, sB).

34. On Star's flying activity, see "Planes Taking to the Lakes," *ADT,* 8 April 1935, 5; and "Little Snow for Landings," *ADT,* 4 April 1935. For Dunkle's nomination to the

Alaska Aeronautical Board, see "Dunkle Urged as Member of Air Board," *FDNM,* 16 April 1937.

35. On the Loussac whiskey incident, see Mills and Phillips, *Sourdough Sky,* 78–83; and DT: R3, t1. The tape states that Dunkle was involved in the search and rescue and that the whiskey was medicinal (and therefore legal, even during Prohibition.) Regarding the fatal crash of Stephen Mills, see Mills and Phillips, *Sourdough Sky,* 147–149.

36. DT: R9, t1, sB.

37. Acting Governor Edward Griffin, telegram to WED, 29 June 1936, Carol Young collection, author's files.

38. WED, letter to Acting Governor Griffin, 6 July 1936, Carol Young collection, author's files.

39. On the construction of the seaway at Lakes Hood and Spenard, refer to DT: R9, t2, sA; and U.S. Department of the Interior, *Annual Report of the Alaska Road Commission,* fiscal year 1940.

40. Mills and Phillips, *Sourdough Sky,* 165; Anchorage Centennial Commission, *Honoring 100 Alaska Bush Pilots.*

41. Information on the Dunkles as an Alaskan flying family comes from William E. Dunkle, conversations with author, 1997–1998; "Mrs. Wesley E. Dunkle Dies in Washington," *ADT,* 30 April 1962; William E. Dunkle, *Birds Eye View,* 298–341; William E. Dunkle, letter to author, 22 April 1998; William E. Dunkle, remarks at a memorial service for John Hull Dunkle, 7 February 1998, St. Patrick's Catholic Church, Carlsbad, Calif.; and Dunkle family interviews, March–April 1997.

42. WED (in Washington, D.C.), letter to Joseph and Stewart Alsop at the *Washington Post,* 1 March 1953. Joseph Alsop answered Dunkle on 10 March 1953: "It is always a pleasure to hear from our readers, and I read what you had to say with a great deal of interest." DF.

CHAPTER 9: A SURE CURE FOR DEPRESSION

A main source for this chapter is a ledger (Ledger A): "Ledger of W. E. Dunkle. Investment Accounts, 1934–1937." The accounts are Airplanes, Anaconda Mining Co., Anchorage Real Estate, Automobiles, Bear Creek 10% Int., Circle Dollar, Dutch Creek, Eva Creek (Fairbanks) mill, Fairview Placer, Golden Horn, Golden Zone, Iditarod Mining Co., Investments Miscellaneous, Loans, Oracle, Pardners Mines, Parks, and Shannon Prospect. The records were almost certainly kept by George M. Campbell, who also managed the sawmill camp operation for Dunkle at the Golden Zone mine. Campbell was also a longtime associate of Arthur A. Shonbeck. This ledger was lent by Bruce B. Dunkle to the Section on History and Archaeology, Alaska Department of Natural Resources, Anchorage.

Ledger A is supplemented by a personal ledger (B) for Earl and Billie Dunkle, which contains records from 1937 to 1940 and includes work sheets for the Dunkle's 1937 federal income tax. It is essentially a check register of their account with the Bank of Alaska (later the National Bank of Alaska, now Wells Fargo). Records include Dunkle's business expenses and personal family records. It was kept by the Dunkles: some years by Billie, other years by Earl.

1. "Nearly Million to Be Spent [on] Matanuska Valley Establishment," *ADT,* 23 April 1935, 1. See also same day and page, "Silver and Tin Are in Spotlight."

2. Peter Bagoy Sr., interview, 1996.

3. Reiss, "Alaska Miner of West Point," 9, 13–14.

4. On Dunkle's early involvement in the Goodpaster region, see "Flying Mining Engineer Tours Interior," *FDNM,* 21 April 1936; Tweiten, *Alaska, Big Delta and the Goodpaster Region,* 37–38, also 84 on Jack McCord; and WED, letter to John Baragwanath, 29 July 1936, Ledger A, Investments Miscellaneous. On the modern Pogo project in the Goodpaster region, see DiMarchi and Friesen, "Exploration Project at Pogo." On the Oracle prospect, see Tuck, "Moose Pass–Hope District," esp. 507–510; and Ledger A, Oracle.

5. Maddren, "Innoko-Iditarod Region," 237–238; Buzzell and Lewis, "Historic Building Survey," 5–10; Buzzell, "Flat and Iditarod," John Miscovich, interview, 22 July 1993, tape 1, side A, 054; Carter, *Iditarod Trail,* 53–54.

6. On the discovery and richness of the Golden Horn lode, see Roehm, "Golden Horn Mine," 8; and Holzheimer, "Otter Creek," 2–7.

7. On the mapping and sampling results, see Dunkle field notebook, 1934.

8. WED, letter to Bert B. Nieding (BBN), undated but ca. June 1934.

9. WED, letter to John Baragwanath, 21 July 1934.

10. On Bert Nieding's employment at the Golden Horn, see WED to BBN, letter and acceptance signature, 30 July 1934. On Dunkle's proposal to Baragwanath regarding the backing of the mine, see WED, telegram to E. B. Kluckhohn, Seattle, 20 October 1934. DF.

11. John Baragwanath to WED, telegram of acceptance from New York, 28 October 1934.

12. On the price of mine timber at the Golden Horn, see BBN, letter to WED, 19 October 1934. On Dunkle's inference of the local high grade of the Golden Horn vein, see WED, letters to BBN, 4 and 27 January 1935. On Nieding's response on the value of the ore and on his development of the vein, see BBN, letter to WED, 13 April 1935.

13. BBN, letter to WED, 16 May 1935.

14. On Nieding's production and local high value of the Golden Horn ore, see BBN, letter to WED, 16 May 1935; and BBN, letters to WED, 1, 2, and 8 June 1935. On Dunkle's visit to the mine, see WED, letter to BBN, 13 June 1935. Author's files.

15. On the value of the Golden Horn ore shipments, see Smelter Settlement Sheets, Lot No. 2398, 6 July 1935, 23.15 tons, and Lot No. 5538, 10 September 1935, 50.91 tons, DF. The lots, respectively, assayed 6.05 and 6.31 ounces of gold per ton and 4.55 and 5.57 ounces of silver per ton. Information on the royalty paid by Dunkle to Justus Johnson comes from the author's conversations with John Miscovich and John Fullerton in 1998.

16. WED, partnership tax return, Golden Horn Mining Co., 1935, author's files; Ledger A, Golden Horn.

17. On expenditures on the Golden Horn prospect, see Ledger A, Golden Horn.

18. The account of Dunkle's venture on lower Willow Creek at Flat through Iditarod Mining Co. has multiple sources, including Ledger A, contemporary newspaper articles, and letters between Dunkle, Baragwanath, and Kluckhohn. It is also based on the author's longtime involvement with Flat. As a young geologist with the USGS in Flat in 1967, I was accepted and brought into its family life by John and Mary Miscovich and their children; Leonard Zaiser and his wife, Eva (née Miscovich); Richard, John, and Jane Fullerton; Alvin and Cathy Agoff; Minnie Brink; and Mr. and Mrs. Henry Stevens. In subsequent years, I promoted projects at Flat and sought a lease on the Golden Horn. I also met other people with a long history at Flat: Andrew

Olson; Charles Awe and his mother, always "Mom"; George and Betty Dahl; Bob and Betty Lyman; and Bob and Rita Sholton. Phil Lindstrom, an ASARCO engineer from Wallace, Idaho, knew quite a bit about Flat because his relative Tony Lindstrom mined there. Too many of these people are now "the late."

Flat is at the "end of the road," although there really is no road to Flat except the short one from Iditarod. Much of the equipment that was brought there is still there. As at other remote mining camps, generations of family miners have learned how to conserve and operate that equipment. Old-timers of Flat ancestry know the operating characteristics and model numbers of equipment brought to the mines in 1934 and before. If Richard Fullerton were still with us, it could be nearly guaranteed that Dunkle's 180-ton P & H dragline would still work and might even be operating. Other less tangible artifacts remain. In 1998 John Fullerton and his son, Tad, still relied on the accurate drill maps prepared by the Guggenheims before 1920. John Fullerton or John Miscovich could recite placer drill records from the early 1900s and knew the men who drilled the holes.

For a long time some Flat miners believed that the big mining outfit of Iditarod Mining Co. was bought and paid for by Sunshine Mining Co. of Idaho. Phil Lindstrom of the Idaho silver belt found Sunshine old-timers who knew that was not the case. The mining equipment was bought by Iditarod Mining Co., that is by Dunkle, Pardners Mines, L. C. Thomson, and E. B. Kluckhohn.

The two persons who helped most on this account were John Miscovich and John Fullerton. They did not always see eye to eye. During the last few years, I "interviewed" both Miscovich and Fullerton several times about past events at Flat. Those interviews, however, fade imperceptibly into conversations held over more than thirty years with the Miscoviches, Fullertons, Agoffs, and Awes, who grew up in the Flat of a much different world.

19. "Manley Mines on Willow to Be Operated," *FDNM,* 21 March 1935, 2.

20. Ibid.; and "Big Dragline to Be Working on Large Placer Gold Area," *ADT,* 11 March 1935.

21. WED, letter to John Baragwanath, 21 July 1934, DF.

22. WED, letter to John Baragwanath, 29 July 1936, DF.

23. Ledger A, Iditarod Mining Co.

24. On Dunkle's investments in the Circle district, see WED, letter to John Baragwanath, 29 July 1936, Ledger A, Circle Dollar. On Dunkle's investments in the Yentna district at Bear Creek and the Fairview placer, see "To Open Up Kuskokwim Cinnabar," *ADT,* 9 April 1935. This article, which describes Dunkle's lease of the Parks Cinnabar property in the Kuskokwim region, also notes exploration activity by the Dunkle crew in the Yentna district. See also Ledger A, Fairview Placer.

25. WED, letter to John Baragwanath, 10 September 1936.

26. On Anaconda's testing of the Shannon ore, see Francis Cameron, letter to M. H. Gidel, Anaconda Co., 27 July 1937, File No. 6520.02, AGDC. See also Dunkle's expenditures on the project, Ledger A, Shannon Prospect.

27. WED, letter to John Baragwanath, 10 September 1936, DF.

28. F. F. Frick, research engineer, Anaconda Co., "Shannon Gold Ore: Report of Preliminary Work," to Reno H. Sales, chief geologist, 3 November 1937, File No. 6520.02, AGDC.

29. Ledger A, Shannon Prospect.

CHAPTER 10: THE GOLDEN ZONE

1. Regarding the Little Annie, see Joralemon, *Adventure Beacons,* 310–311. On the character of the ore at Shannon's Slippery Creek deposit, see notes 26 and 28 in chapter 9, "A Sure Cure for Depression."

2. See note 47 in chapter 6, "From Pier 2, Port of Seattle, to Africa." In 1931, when Billie planned to leave the University of Cape Town, she received a recommendation from her former professor, Lancelot Hogben, who was then at the London School of Economics and Social Work. In the letter (15 January 1931), Hogben wrote, "[S]he demonstrated with great zeal and efficiency in my own laboratory to large classes of medical students during the session of 1930." Billie was raised at Crookston House by Alfred E. Borthwick, Scottish artist and inventor ("Death of Noted Artist, Captain Alfred E. Borthwick," *The Scotsman,* 8 December 1955), who probably was Billie's uncle. Billie received a British patent (no. 417793), which was granted on 12 April 1933 for one invention; in later years she corresponded with her husband about other inventions.

3. On the USGS's investigations of the Chulitna (Golden Zone) area, see Ross, "West Fork of the Chulitna River," 291. On the ASARCO exploration program, see Hawley and Clark, *Upper Chulitna District,* B33. See also John Balla, ASARCO, letter to author, 1 April 1996.

4. "Flying Mining Engineer Tours Interior," *FDNM,* 21 April 1936.

5. Ledger A, Golden Zone. The first payment to Lon Wells (five hundred dollars) was made on 9 August 1935.

6. Letter and metallurgical report to Henry C. Carlisle, San Francisco, Calif., from Minerals Separation North American Corp., transmittal letter by Edward Nutter, chief engineer, 5 February 1936, and report by Carl F. Williams, same date, author's files. Williams stated, "The results of test 4 show, however, that flotation concentration followed by cyanidation of reground—but unroasted—flotation concentrates would be productive of good results."

7. WED, letter to John Baragwanath, 29 July 1936, DF.

8. U.S. Department of the Interior, *Annual Report of the Alaska Road Commission,* fiscal year 1936, 17–18. See also U.S. Department of the Interior, *Annual Report of the Alaska Road Commission,* fiscal year 1937. These reports are available from the Alaska Department of Transportation and Public Facilities, Anchorage.

9. Peter Bagoy Sr., interview, 1996.

10. J. F. Bowes, England, letter to WED, 13 March 1938, DF.

11. Francis Cameron, letter to Murl H. Gidel, Butte, Mont., 27 July 1937, File No. 6520.02, AGDC.

12. Francis Cameron, "Summary and Preliminary Report," letter report to Reno H. Sales, chief geologist, Anaconda Co., New York, 25 September 1937, 6–7, File No. 6520.01, AGDC.

13. Francis Cameron, telegram to WED, 25 February 1938; Francis Cameron, letters to Reno H. Sales, 23 and 26 February 1938; File No. 6520.02, AGDC.

14. WED, letter to Francis Cameron, 25 February 1938, File No. 6520.02, AGDC.

15. Francis Cameron, letter to WED, 11 March 1938, File No. 6520.02, AGDC; and author's files.

16. Ibid. See also WED, letter to Francis Cameron, 25 February 1938, File No. 6520.02, AGDC.

17. Golden Zone Mine, Inc. corporate documents, including the subscription agreement of 12 May 1938 and the mine lease of 14 May 1938, author's files.

18. On Alex McDonald, see "Funeral Conducted for Alaska Pioneer [Alex McDonald]," *ADT,* November 1962. On Tarwater, see "E. R. Tarwater Dies in Seattle," *ADT,* 7 August 1944. Tarwater was also active in local public affairs and was appointed as a strike arbitrator in a dispute over school construction; see "Tarwater Named Labor Arbiter," *ADT,* 6 July 1938.

19. On A. A. Shonbeck, see "Shonbeck Rites Are Set for Next Sunday," *ADT,* 22 June 1945. See also Shonbeck's application for membership, Benevolent and Protective Order of Elks, Anchorage, ca. 1918, Elks Club files, Anchorage; and Evangeline Atwood, *Anchorage: All-American City,* author's note, v.

20. Data from shareholder list, Golden Zone Mine, Inc., 1945, DF.

21. Information on investors in Golden Zone Mine, Inc. comes from a collection of Bank of Alaska (later National Bank of Alaska; now Wells Fargo) deposit slips that name the depositors and the amounts deposited. DF.

22. WED, letter to Harry Townsend, 22 March 1939, File No. 6520.02, AGDC.

23. "Golden Zone Hits $68 Ore in New Vein," *ADT,* 20 July 1939; "Fine Vein of Ore Struck in Golden Zone," *FDNM,* 12 July 1939; "New Mines Development Significant," editorial, *FDNM,* 14 July 1939.

24. Letter report, Tacoma smelter to WED, 9 August 1939, author's files.

25. "Fishing and Mines Theme at C. C. Meeting," *FDNM,* 14 August 1939, 3.

26. WED, "Golden Zone Mine: Report to Shareholders."

27. WED, letter to Harry Townsend, Seattle, 22 March 1941, File No. 6520.02, AGDC.

28. Flow sheet filed with shareholders report (see note 26 above). See also Alexander Smith, "Golden Zone Mine."

29. Alexander Smith, "Golden Zone Mine."

30. WED, letter to Harry Townsend, Seattle, 22 March 1941, File No. 6520.02, AGDC.

31. Ibid.

32. Kennecott Copper Corp., "Annual Report for 1938"; Moerlein, "Golden Zone Property Evaluation." Also refer to Dunkle tapes.

33. WED, letter to Harry Townsend, Seattle, 22 March 1941, File No. 6520.02, AGDC.

34. Ibid.

35. On prospecting privileges extended to workers at the Golden Zone mine, refer to the Paul Ellis interview, 1996. On the Golden Zone crew, many of whom were from the Lucky Shot mine, see "Crew of 30 at Work on Quartz Job," *FDNM,* 4 August 1938, 7. The article quotes miner Anton Covich: "The men at the Golden Zone comprise a picked crew who formerly worked for Mr. Dunkle at the Lucky Shot." On the financial interest of the miners at the Golden Zone mine, see Kennedy, "Forty Men," 8.

36. Luther ("Tex") Noey, interview, 1996.

37. WED, "Golden Zone Mine: Report to Shareholders."

38. C. F. Herbert's opinion on RFC financing was surmised from Herbert's conversation with the author on November 1998. On a 1940 appeal by Dunkle to Baragwanath for project funding, see WED, letter to J. G. Baragwanath, Chrysler

Building, New York, 9 April 1940, DF. On the impending war, see Winifred Heitmeyer, letter to GBD, 15 October 1940, DF.

39. Employee names and statistics come from the Golden Zone payroll book, 1940–1941, Carol Young collection, author's files. On the first milling at the mine, see "Golden Zone Crushes Its First Ore," *FDNM,* 8 October 1940, 7; and "One Shift Is Operating at Golden Zone," *FDNM,* 9 January 1941, 3.

40. The USGS reported 869 tons of concentrates that contained 1,581 ounces of gold, 8,617 ounces of silver, 21 tons of copper, and about 3,000 pounds of lead (Hawley and Clark, *Upper Chulitna District,* B34). See also Alexander Smith, "Golden Zone Mine," table of smelter returns.

41. The Tacoma smelter penalized ore shipments that contained between 1 and 17 percent arsenic. If it was below 1 percent, arsenic was a tolerable nuisance. If the arsenic concentration was between 1 and 17 percent, its recovery costs would be higher than its sale profits. With an arsenic concentration above 17 percent, arsenic sales made the smelter a small profit, but that profit was not shared with the mine. At least the Golden Zone ore was not penalized.

CHAPTER 11: PLACER GOLD, COAL, AND POSTWAR REALITY

1. On the history of development in the Kantishna mining district, see Buzzell, "Caribou Creek Drainage History," 4–7. On prospecting for W. P. Hammon, see Bigelow, "Caribou Creek."

2. See WED, letters to Glenn Carrington (GC), 11 October 1938 and 16 November 1938, DF, on lease agreements and payments to Maurer and Meehling, Jas. S. Newlan, and Bill Taylor and on the lease of placer ground owned by Mt. McKinley Placers. See also GC, letter to WED, 4 October 1937, DF. Further details on the formation of the company are in E. B. Kluckhohn, letter to WED, 21 April 1939, DF. The total purchase price for the property was $135,500, of which $13,000 was to be in cash and the balance in royalty. L. C. Thomson agreed to be a partner in a telegram he sent to WED on 9 December 1938 (DF).

3. GC, letter to WED, 4 October 1937, DF; Arthur F. Erickson (for GC), letter to WED enclosing $750 for prospecting share, 11 July 1938, DF.

4. Arthur F. Erickson, letter to WED, 25 November 1938, DF; Erickson, "Caribou Creek," 4; GC, letter to WED, 9 December 1938, DF.

5. Arthur F. Erickson, letter to WED, 16 December 1938, DF. A second letter to WED from Erickson on the same date gives operating details on the washing plant.

6. GC, letter to George M. Campbell, 3 April 1939, DF. Campbell kept the financial records for both the Golden Zone and the Caribou Creek projects in his office at the Golden Zone mine.

7. "W. E. Dunkle Developing Two (2) Properties," *FDNM,* 17 April 1939, 7; "Dragline to Be Taken to Dunkle Mine," *FDNM,* 21 February 1939, 7.

8. T. S. Emerson, manager, Alaska Brokerage Co., letter to GC, 31 March 1939, DF.

9. See note 7, this chapter. On the hiring of skilled employees by Carrington, see GC, letters to WED, 3 March 1939 and 21 April 1939, DF.

10. Memorandum report on (gold/silver) bullion deposited at the U.S. Assay Office, Seattle, No. 308, Bar 404, 18 July 1939, DF; also memo report, as above, on second gold bar, 27 July 1939. The second bar contained 199.2 ounces of fine gold.

Details on project gross and expenditures are given in an unsigned memorandum titled "1939 Operations of Caribou Mines." The memo was filed with a handwritten letter from George M. Campbell, 29 October 1939, and was probably written by Campbell (DF).

11. Details on workers' compensation and other insurance are given in a letter from E. B. Kluckhohn to WED, 7 April 1939, DF.

12. On Medford's accident, see George M. Campbell, letter to E. B. Kluckhohn, 10 July 1939, DF. On Fogelman's probably fatal shooting, see WED, letter to Jack Turnbull, 17 September 1939, DF; and George M. Campbell, letter to E. B. Kluckhohn, 7 February 1940, DF.

13. GC, Seattle, letter to WED, 29 March 1939, DF.

14. Ibid.

15. Gold production data for Caribou Mines Co. were compiled from U.S. Bureau of Mines records on the Kantishna district: "Production and Development of Metal Mines." Production data were submitted annually by the operator (Caribou Mines Co.) to the Bureau. These records are now in the National Archives, Anchorage. Further information on the productivity of the mines is given in "Caribou Mines Uncovers Million Square Feet in Season's Placer Work," *FDNM,* 26 September 1940, 5; and Arthur F. Erickson, letter to Thomas K. Bundtzen, ADGGS, 25 October 1976, author's files.

16. On Carrington's view of the postwar reopening of Caribou Mines Co., see GC, letter to WED, 19 March 1949, DF. On Dunkle's sale of his interest in Caribou Mines Co. to Carrington, see WED, letter to GC, 13 March 1946, DF.

17. Regarding wartime contributions made by officials of Pardners Mines, Harold E. Talbott wrote the following to WED on 10 January 1946 (DF): "When the war broke out Phil Wilson went with the War Production Board in the metals division and Jack [Baragwanath] went off to operate a nickel property in Cuba we had become interested in."

18. On the wartime contributions made by other geologists and engineers in Dunkle's circle, see note 1 in chapter 7, "Pardners Mines and the Lucky Shot"; "Harold E. Talbott," in *Who Was Who in America,* vol. 4 (1942); and Joralemon, *Adventure Beacons,* 366–367. On Levensaler, refer to DT: R5, t1, sB.

19. On the discovery and initial use of the Costello Creek (Dunkle Hills) coal, see Capps, "Upper Chulitna Region," 231–232. On subsequent developments, see Rutledge, *Dunkle Coal Mine,* 3–4.

20. DT: R5, t1.

21. On Billie Dunkle's wartime contributions and interests, see "News and Previews From the U.S.O.," *ADT,* 27 January 1943; and Edd Johnson, chief, Control Division, Office of War Information, New York, letter to GBD, 23 July 1942, DF. Regarding the value of Dunkle's coal mine, see GBD, draft letter to Raymond Davis, U.S. Department of the Interior, Washington, D.C., undated but ca. 1953, DF. Regarding her husband's coal production records, Billie wrote,

> During the war he supplied coal to the Army during a period when the fuel situation became so tight that both Elmendorf and Ladd bases were down at times to only a few hours supply. Unfortunately, the situation developed so rapidly that it was not possible to increase production to an adequate point, but the difference in price between his coal at eleven

dollars per ton as against a price of forty-five dollars for coal which had to be imported from the States saved the Government many thousands of dollars, at the same time saving critically needed shipping space.

22. About Alexander Smith, see CIM, "A. Smith." The following correspondence concerns the examination of the Golden Zone mine: Alexander Smith, Fairbanks, letter to WED, 4 September 1944, DF; Herbert J. Waugh, Juneau, letter to WED, 11 August 1944, DF; WED, letter to Herbert J. Waugh, 24 August 1944, DF; and WED, letter to Alexander Smith, 8 September 1944, DF.

23. Alexander Smith, "Golden Zone Mine," esp. 27–28.

24. Ibid., 2.

25. This paragraph and succeeding ones are largely based on company correspondence, especially a file with more than fifty letters and telegrams concerning the possible acquisition of the Golden Zone mine by St. Eugene Mining Co. St. Eugene was a subsidiary of Ventures, Ltd., a company run by one of the great Canadian mining entrepreneurs, Thayer Lindsley. Lindsley's representatives on the Golden Zone transaction were Alexander Smith and Ridgeway R. Wilson. The main parties on Golden Zone's side were Dunkle, Glenn Carrington, and Arthur A. Shonbeck. Edward A. Rasmuson of the Bank of Alaska was an important contributing player. The file covers a period from August 1944 until mid-1946; author's files. Details on the initial negotiations are given in the following: GC (in Seattle or Vancouver, Canada), telegram, to WED, 31 October 1944; WED, telegram to GC, 1 November 1944; and WED, letter to GC (cc: Ridgeway R. Wilson [RRW]), 1 November 1944.

26. GC, letter to WED (cc: Edward A. Rasmuson and Alexander H. McDonald), 2 November 1944; GC, letter to WED (cc: Elmer E. Rasmuson and Alexander H. McDonald), 7 November 1944; and Arthur A. Shonbeck (AAS), letter to WED, 15 November 1944.

27. WED, letter to GC, 16 November 1944.

28. Chattel Mortgage, Golden Zone Mine–Bank of Alaska, 14 January 1944, author's files.

29. DT: R9, t2; WED, letter to GC, 16 November 1944; Alexander Smith, "Golden Zone Mine," 2.

30. WED, letter to Miriam Dickey, secretary to Cap Lathrop, publisher of the *Fairbanks Daily News-Miner,* 26 December 1944; WED, letter to Robert Atwood, publisher of the *Anchorage Daily Times,* 27 December 1944; "Sale Made of Golden Zone Mine," *FDNM,* 28 December 1944; "Mines to Reopen This Year," *FDNM,* 8 January 1945.

31. On Edward Rasmuson's view of Golden Zone's creditors, see Edward A. Rasmuson, letter to GC, 26 December 1944. On Carrington's and Shonbeck's fundamental agreement with this position, see GC, letter to Edward A. Rasmuson (cc: WED), 3 January 1945. About the Christmastime train wreck, see GC, letter to WED, 5 January 1945; and RRW, letter to WED, 9 January 1945.

32. WED, letter to J. P. Garvin, attorney, Seattle, 12 February 1945; J. P. Garvin, letter to WED, with reference to a Kluckhohn wire and research on tax matters, 25 January 1945.

33. Alexander Smith, letter to WED, 13 February 1945.

34. GC, letter to WED, 19 February 1945; AAS, telegram to WED, 19 February 1945. Shonbeck suggests that they let St. Eugene settle tax matters and that Dunkle, Edward A. Rasmuson, Carrington, and Shonbeck meet in Seattle in early March 1945.

35. WED, transmittal letter to AAS, 1 March 1945, and memorandum with three main topics: (1) a memorandum to GZM creditors and stockholders; (2) the standby agreement regarding creditors and stockholders; and (3) a sinking fund to be used for repair and maintenance of equipment. The sinking fund and the use and maintenance of equipment are extensively discussed in a letter from WED to GC, 30 January 1945.

36. RRW, letter to WED, 2 March 1945; WED, letter to GC (cc: AAS), 9 March 1945. Quote from WED, letter to AAS, 10 March 1945.

37. Golden Zone Mine, Inc. (WED), letter to creditors and stockholders, 22 March 1945. This letter outlines the St. Eugene deal and includes an attached copy of a new draft of the standby agreement.

38. RRW, letter to WED, 15 June 1945.

39. On postwar wages and progress on the standby agreement, see WED, letter to RRW, 29 June 1945. Regarding direct negotiations between the Bank of Alaska and St. Eugene, see WED, letter to RRW, 8 August 1945.

40. "Shonbeck Rites Are Set for Next Sunday," *ADT,* 21 June 1945.

41. WED, letter to GC, 2 March 1948, author's files.

42. Regarding the Golden Zone property, see R. F. Mahoney, Callahan Zinc-Lead Co., Inc., Wallace, Idaho, letter to WED, 16 June 1948. Related correspondence includes the following: H. C. Gunning, New Jersey Zinc Exploration Co., Ltd., letter to WED, 20 May 1948; and E. A. Julian, vice president, Goldfield Consolidated Mines Co., San Francisco, letter to WED, 27 May 1947. Private contacts included E. G. Frawley, attorney, Salt Lake City (WED letters of 26 and 29 August 1948); and contacts made through Carrington (WED, letter to GC, 30 November 1946). Contacts with Alaska-Juneau Mining Co. include J. A. Williams, general manager, letter to WED, 28 November 1947; WED, letter to Alaska-Juneau Mining Co., 22 November 1947; and WED, letter to H. L. Faulkner, Esq., 22 November 1947. Williams forwarded Dunkle's correspondence to the head office of the company in San Francisco, but no reply exists in the files. Author's files.

43. WED, "Golden Zone Mine, Inc."

44. Minutes, Board of Directors, Golden Zone Mine, Inc., 22 October 1948, at the law office of Warren N. Cuddy, Anchorage. On the sale of assets to Usibelli, see Bill of Sale, Golden Zone Mine to Usibelli Coal Mine; and GC, letter to WED, 17 August 1949. The sale took place before 15 October 1949; the disposition of funds was described in a letter from WED to Carrington and Co. on 17 March 1950. Author's files.

45. On the Snow Bird prospect, see WED, letter to GC, 14 December 1948. Regarding Newmont's gold-mining operations, see DeWitt Smith, letter to WED, 21 May 1948. Author's files.

46. Earl Dunkle and his oldest son, John, mined at the Costello Creek coal mine in the summer of 1947. At the very least, they must have formed the most highly educated team of coal miners in Alaska. The father graduated from Sheffield Scientific School. The son had an M.A. degree from St. Andrews University, had studied at Oxford, and had earned law and accounting degrees from New York University. On the retrieval of grinding balls from Golden Zone, see WED, letter to Elmer E. Rasmuson, 17 October 1950, author's files.

47. GBD, draft letter to Raymond Davis; see note 21, this chapter. See also War Department, Notification of Personnel Action, Headquarters, U.S. Army–Alaska,

APO 942, Seattle, 2 April 1951, which mentions the promotion from engineering draftsman (GS-5) to mechanical engineer (GS-9). Author's files.

48. On Billie's popcorn and sketching ventures, see David Thorsness, Anchorage, conversation with author, 1996; BBD, interview with author, April 1997; Alice Johnson, Seattle, letter to GBD, 14 February 1942, DF; and "Seattle Paper Lauds Scenes by Alaskan," bylined *Seattle Post-Intelligencer,* undated but ca. spring 1942, DF.

49. Regarding Z. E. Eagleston, see Renkert and others, "Weimer, Robert E. (Pete) and Judy," 39; Kent Woodman, conversation with author, 1997; and Robert B. Atwood, "Between Us." The information on Eagleston is also based on the Paul Ellis interview, 1996; the author's conversations with Herb Rhodes, 2000; Z. E. Eagleston, Steilacoom, Wash., letters to GBD, 17 July 1949, 14 and 21 August 1949, DF; and the views of Maxine Reed (Mrs. Frank Reed Jr.), conversations with author, 1997 and 1999.

50. Felix D. Logsdon, Office of the 942-2 Branch Exchange, Elmendorf Air Force Base, Alaska, letters to GBD (in Washington, D.C.), 18 July 1952, 30 March 1953, DF; Dunkle family interviews, 1997.

51. John Baragwanath, letter to GBD, 11 June 1952, DF; John S. Minary, Shelter Rock Development Corp., letter to Brig. Gen. Robert A. McClure, 11 June 1952, DF. Minary was William Paley's longtime factotum and a partner of Baragwanath in Shelter Rock Development Corp.

52. WED, letter to Charles Bunnell, president, University of Alaska–Fairbanks, 23 February 1953, DF; WED, telegram to Harold Talbott, Secretary of the Air Force, Pentagon, 19 February 1953, DF; WED, letter to Andrew Nerland, Fairbanks, 23 February 1953, DF.

53. WED, letter to Andrew Nerland, Fairbanks, 23 February 1953, DF.

54. WED, letter to Charles E. Bunnell, College, Alaska, 23 February 1953, DF.

55. Andrew Nerland, letter to WED, Washington, D.C., 5 March 1953, DF; WED, letter to GC, 7 March 1953, DF.

CHAPTER 12: ONE MORE EXPERIMENT

1. WED, letter to Bradford Washburn, 23 March 1953, DF.

2. Ibid.

3. Bradford Washburn, letter to WED, 30 March 1953, DF.

4. Howard Cady, letter to Bradford Washburn, 6 April 1953, DF.

5. On the taping sessions in Washington, D.C., refer to the Dunkle tapes, esp. DT: R1, t1 and t3, and DT: R2, t2; and BBD, Dunkle family interviews, 1997. On the contributions of L. A. Levensaler to Dunkle's knowledge of Copper River history, see WED, letter to "Lewis" Levensaler, 29 June 1953, Levensaler files, APRD. Dunkle acknowledges a letter from Levensaler: "Thanks a lot for your long letter about early days along the Copper River. Not only will it help a lot in my trying to get that piece of work done but I read it all with great interest, because it added a great deal to my knowledge of those times." On Levensaler and Henry Watkins, refer to DT: R5, t1, sB.

6. Several people have proposed that a book or article on Helen Van Campen would be valuable. Journalist Kay J. Kennedy of Fairbanks quotes Bob Davis, former publisher of the *New York Sun* and the Munsey Publications: "Tell Helen Green that if she'll write the story of her life I'll publish it" (Kennedy, "Helen Van Campen Dies"). George A. Thompson Jr. of New York University proposed to write an article about

Helen in 1991: "I am hoping to write an article about her which will lift her from total oblivion into mere obscurity." Thompson wrote Dan Fleming, curator, Alaska collection, Anchorage Municipal Libraries, on 16 May 1991, Alaska section, Z. J. Loussac Public Library, Anchorage. Renee Blahuta, a Fairbanks archivist who has Kay Kennedy's papers, has also expressed an interest in writing about Helen. Also refer to the Dunkle tapes, esp. DT: R5, t2. For a not-too-flattering assessment of Baragwanath, see Sally Bedell Smith, *In All His Glory,* esp. 109–110, 324–325, and 458. On Dunkle's promise to Cady, see WED, letter to Howard Cady, 11 April 1953, DF.

7. Apell, "Broad Pass Coal Reports."

8. Douglas Colp, mining engineer, Fairbanks, conversation with the author, 10 March 1996.

9. WED, letter to Mrs. Ferrick (mother of Dorothy [Mrs. Jack] Dunkle), 11 January 1957, DF.

10. On Dunkle's leases and activity reports on the Broad Pass and Middle Fork coalfields, see Federal Form 374-D, monthly report on activities on Permit No. 027608 (Middle Fork) from August 1954 until June 1956, U.S. Bureau of Land Management, DF. See also Mrs. Margery J. McCormick, acting manager, U.S. Bureau of Land Management, letter to WED, 23 August 1954, DF. The letter concerns Coal Prospecting Permit No. 011636 (Broad Pass); it quotes a USGS determination on the status of the site. See also U.S. Bureau of Land Management Permit No. 011636, Broad Pass Prospecting Permit, issued 7 September 1954; and U.S. Bureau of Land Management Permit No. 027608, Middle Fork Prospecting Permit, ca. spring 1954, application for amendment filed in 1955; DF.

11. C. F. Herbert Jr., report on Broad Pass coal, cited in Accolade Mines, Inc., "Consolidated Report, 1961."

12. WED, letter to BBD, 22 November 1955, DF.

13. WED, letter to BBD, 28 October 1956, DF.

14. On extreme weather in Broad Pass, see WED, letter to BBD, 4 November 1956, DF; WED, letter to BBD, 5 January 1957, DF; Peter Bagoy Sr., interview, 1996; and WED, letter to BBD, 15 February 1957, DF. Earl wrote to his son:

> The building in the picture, which you will recognize as the old warehouse, is where I'm setting up to do the drying. The coal is on the ground behind it, buried in several feet of snow, but I have to dig out only a little over a ton a day. . . . [I]t will not be too hard, although I have to heave it well over my head with the shovel in order to get through the back door and onto the floor. I am living in the room at the south end of the building and find it easy to keep warm, because it is very tight and well insulated. When I get the boiler going steady in the main room the whole building will be warm. At present that other room makes me a wonderful "walk-in refrigerator" twenty by thirty feet and well below zero.

On the delay in obtaining the coal-drying boiler, see WED, letter to BBD, 26 January 1957, DF.

15. Company document, Alaska Exploration and Development Co., 1954, DF.

16. GBD (in Washington, D.C.), letter to BBD, 15 September 1955, DF.

17. WED, letters to BBD, 28 October 1956, 4 and 20 November 1956, and 2 December 1956, DF.

18. GBD, letter to BBD, 18 August 1955, DF. Billie Dunkle's role in the Eisenhower campaign was also noted in other correspondence: Gladys Brooks, chair, "Citizens for Eisenhower," Minneapolis, letter to GBD, 30 October 1956; and Mrs. Stanley (Sue) Moore, Fourteenth Congressional District, "Citizens for Eisenhower—Southern California," Los Angeles, letter to GBD, undated but pre-November 1956; DF. Billie's role was also mentioned in a press release: "Portrait of Mrs. Eisenhower Presented to Her on Her Birthday (Today) by Women Leaders of National Citizens for Eisenhower-Nixon," 1956 Citizens for Eisenhower-Nixon, Washington, D.C., 4 November 1956, DF. The release states, "The painting was presented today by Mrs. W. E. Dunkle, Deputy Chairman, Women's Finance."

Billie also served as secretary of the Washington State and Alaska Society, as mentioned in the program for the annual meeting of the society, held on 23 April 1956 in the National Press Building, Washington, D.C. (DF). She was also secretary of the Washington Home Rule Committee, as cited in the following documents: Secretarial notebook labeled "Home Rule," Special Events Subcommittee, Mrs. J. Borden Harriman, chair, first meeting 18 August 1954, subsequent meetings through 23 November 1954; typed minutes, 27 May 1955; and program, Benefit Ball, Washington Home Rule Committee, Presidential Room, Statler Hotel, Washington, D.C., 14 May 1955; DF. Both Billie and her associate J. Bernard McDonnell are listed on the Ball Committee.

19. On Dunkle's animal friends at Broad Pass, see WED, letters to BBD, 18 October 1956, 26 January 1957, 5 February 1957, DF.

20. On the Kirsch family, Dunkle, and friends at the Broad Pass section house of the Alaska Railroad, see WED, letter to BBD, 5 February 1957, DF; WED, letter to GBD, 5 February 1957, DF; and telephone conversation between Rose Kirsch, Tenn., and author, 18 April 1998.

21. Appointment letter, Arts and Crafts Board, U.S. Department of the Interior, to GBD from Fred H. Massey, U.S. Department of the Interior, undated but ca. spring 1957, DF. A photograph of Billie in her Arts and Crafts role appears in *FDNM,* 4 September 1957. See also notes for a talk on the Native Arts and Handicrafts Program, ca. September 1957, DF; and GBD, letter to Governor Michael A. Stepovich, Juneau, 12 June 1957, DF.

22. On the importance of the Arts and Crafts Program, see WED, letter to BBD, 21 July 1957, DF. On the sale of partial interest in the coal project, see WED, letter to GBD, December 1956, DF. The letter notes the sale of a 6 percent interest in the coal mine project to Goodnews Bay Mining Co. Goodnews Bay may have acquired more of the project after Dunkle's death. Other correspondence on Goodnews Bay's involvement in the project include GBD, letters to BBD, 17 and 19 October 1957, 8 November 1957, and 10 December 1957, DF.

23. U.S. Department of the Interior, agenda, Indian Arts and Crafts Board Meeting, scheduled for 27 and 28 September in room 4004, Interior Building, Washington, D.C., DF. See also note 21, this chapter. The quote is from "Summary—Value in Dynamic Arts and Crafts Program," by GBD, 26 September 1957, DF.

24. WED, letters to BBD, 22 and 27 September 1957, DF.

25. On the search for Dunkle, see "W. E. Dunkle Missing in Wilderness," *ADT,* 3 October 1957, 1, 9; "Dunkle Search Widens," *ADT,* 4 October 1957, 1; "Searchers Find Dunkle's Body," *ADT,* 5 October 1957; "Plane With Four Aboard Missing in

Alaska Range," *FDNM,* 5 October 1957, 1; "Geologist's Body Found Near Stream," *FDNM,* 8 October 1957, 7; and author's telephone conversation with Dan Cuddy, March 1999.

26. On Dunkle's funeral and survivors, see "Dunkle Rites Set Tuesday," *ADT,* 7 October 1957, 9. On the views about Dunkle held by Sheffield's 1908 classmates, see H. DeWitt Smith, New York, letter to GBD, 9 December 1957, DF; and Howard Oliver, Letter in "1908 Class Notes."

27. March, "Report of Coal Drying." See also Accolade Mines, Inc., "Consolidated Report 1961"; the cover also states, "Private Placement; $350,000 Convertible Note."

28. Accolade Mines, Inc., "Consolidated Report 1961"; Edward T. McNally, Pittsburg, Kans., letter to author, 7 March 1997. See also GBD, letter to Governor William A. Egan, 5 December 1958, DF, which quotes Billie: "We will start building our processing plant early next spring. At present one is being built by Truax-Traer in North Dakota for their lignites which closely resemble those of the Broad Pass area in Alaska."

29. On the Alaska Jupiter claims, see GBD, letter to BBD, 28 June 1958, DF. On Billie's membership in AIME, see Ernest Kirkendall, secretary of AIME, New York, letter to GBD, 29 November 1958, DF: "It is a privilege to advise you that you have been elected to the membership in the AIME and to welcome you."

30. "Mrs. Wesley E. Dunkle Dies in Washington," *ADT,* 30 April 1962.

31. Ibid.

EPILOGUE

1. Wickersham, Introduction to *Alaska,* 17. See also the note on Nichols, note 23, chapter 2, "Alaska in the Good Years."

2. Janson, *Copper Spike,* 107–117.

3. WED, "Copper River District," 2.

4. DT: R8, t1, sA.

5. "Origin and Inception of the Heavy Metals Program," internal memorandum, U.S. Bureau of Mines and USGS, ca. 1968, 42 p.; memo sent 24 May 1991 by Chuck Hoyt, U.S. Bureau of Mines, Washington, D.C., to author. In October 1965, President Johnson's science advisor, Dr. Donald F. Hornig, expressed the administration's concern about the gold drain and related problems to T. F. Bates, science advisor to the U.S. Department of the Interior, and to William T. Pecora, director of the USGS. Further discussions were held on 2 November 1965. The meeting of a blue-ribbon panel of scientific advisors, chaired by Hornig, was held on 3 March 1966. The scientists recommended a five-year program to be carried out by the USGS and the U.S. Bureau of Mines. In March 1966, $10 million was reprogrammed within the U.S. Department of the Interior and $1 million was added by the U.S. Department of the Treasury for the Heavy Metals Program, which sought new domestic sources of gold.

6. Baragwanath, *Good Time Was Had,* 90.

Glossary of Mining and Geologic Terms

adit. *See under* **mine workings.**

amalgamation. A mineral-recovery process that involves forming an alloy between mercury, a liquid at room temperature, and another metal, especially gold. The process has largely been superseded by *cyanidation,* but it was used extensively in the Willow Creek district and elsewhere in Alaska before World War II to recover native gold. In early practice in the Willow Creek district, a mercury-coated copper plate below a stamp mill seized the free gold in the mill pulp that flowed over the plate. At the Lucky Shot mine, Dunkle's mill man, Leo J. Till, used a rotating amalgam pan to chemically seize the gold particles. The gold and mercury were separated by distillation.

anticline. A fold of stratified rocks whose inner layers contain the older rocks; it is generally convex upward. *Contrast* **syncline.**

bedding. "The arrangement of a sedimentary rock in beds or layers of varying thickness and character; the general physical and structural character of the beds and their contacts within a rock mass" (Bates and Jackson, *Glossary of Geology,* 59).

bedding plane. "A plane or nearly planar bedding surface that visibly separates each successive layer of stratified rock . . . from the preceding or following layer" (Bates and Jackson, *Glossary of Geology,* 59–60).

bedding-plane fault (or bedding fault). "A fault whose surface is parallel to the bedding planes of the constituent rocks" (Bates and Jackson, *Glossary of Geology,* 59). At Kennicott the miners called the prominent fault at the base of the main ore bodies the *Bedding Plane* (WED, "Copper River District," 10).

bedrock foot. The total amount of gold in one square foot of a placer deposit. It is an alternative way of stating the grade or richness of a placer deposit (as contrasted with ounces or grams per cubic yard or meter). The bedrock-foot system is often favored by placer miners, who can visualize the gold falling to the bottom of the pay streak, generally the top of bedrock. It is also easily derived from the amount of gold recovered in a placer cut measured in square feet (the length times the width of the cut).

breccia. "A coarse-grained . . . rock composed of angular broken rock fragments held together by a mineral cement or [embedded] in a fine-grained matrix" (Bates and Jackson, *Glossary of Geology,* 81). The breccias described in this book formed either by faulting or by collapse into a structure created through hydrothermal leaching.

concentrate. The product of the milling process, in which ore is crushed, ground, and concentrated by gravity or physical processes, including *flotation.* Concentration is effected by removing the *gangue,* the valueless and usually less dense rock material incorporated with the ore minerals. The concentrate may be treated by chemical, electrochemical, or pyrochemical processes to produce pure metals.

crosscut. *See under* **mine workings.**

cutoff grade. In ore estimation, the lowest grade that will meet the costs of mining and milling, not including the repayment of capital.

cyanidation. A mineral-recovery process that involves dissolving native gold in a weak solution of cyanide salts (NaCN or KCN) and then extracting that gold by depositing it on metallic zinc (Merrill-Crowe process) or on activated charcoal. (In the latter method, the gold is then stripped from the carbon.) The gold is cast into impure (Doré) bars and is further purified chemically or electrochemically.

decline (or incline). *See under* **mine workings.**

depletion allowance. "A proportion of income derived from mining or oil production that is considered to be a return of capital not subject to income tax. It is a way of recognizing that mining or oil production ultimately exhausts the reserve" (Bates and Jackson, *Glossary of Geology,* 167).

diagenetic deposit. A deposit formed during the process of *diagenesis,* the physical, chemical, and biological modifications that occur in a sediment after it is deposited and during and immediately after it lithifies. This term may also describe the early modifications to a volcanic or volcanogenic sediment. *See also* **epigenetic deposit** and **syngenetic deposit.**

dike. A tabular igneous intrusion emplaced on a structural plane that cuts across the bedding or layering of the rock. *Contrast* **sill.**

dip. "The angle that a structural surface, e.g., a bedding or fault plane, makes with the horizontal, measured perpendicular to the *strike* of the structure and in the vertical plane" (Bates and Jackson, *Glossary of Geology,* 176). *See also* **strike.**

drift. *See under* **mine workings.**

epigenetic deposit. A mineral deposit formed later than its host rock. The deposits at Kennicott, Lucky Shot, and Golden Zone are epigenetic. At Kennicott, the limestone and dolomite hosts are Triassic in age; the mineral deposits are probably latest Cretaceous or earliest Tertiary, a gap of about 100 million years. At Lucky Shot and Golden Zone, the time gap is not as long, but at both places intrusive host rocks cooled and were faulted before the mineral deposits formed. *See also* **diagenetic deposit** and **syngenetic deposit.**

flotation. A mineral-separation process in which finely ground minerals float in an aerated water-based froth, whereas other minerals sink. Reagents are added to the water that are specific to the mineral being isolated.

flux. A mineral substance that promotes one or more of the following processes in smelting: fusion, fluidity (in a copper sulfide matte, for example), or waste removal in slag that floats on the matte.

footwall. The wall or rock under the ore; the underside of the ore in relation to the dip of the ore body. *Contrast* **hanging wall.**

gangue. The valueless rock or mineral aggregate in an ore that cannot be avoided in mining. It is removed during the milling (concentration) process.

glory hole. *See under* **mine workings.**

gossan. *See* **leached cap or capping.**

grade (of ore). The average assay of a tonnage of ore. Grades of copper and base metal ores are commonly given in weight percent of metal. Grades of precious metal ores are given in ounces per ton or, in the metric system, in grams per metric ton. Dunkle and many other U.S. engineers of his time often computed the grade of gold deposits in dollars per ton relative to a government-fixed price of metal, either the $20.67-per-ounce price before 1934 or the $35.00-per-ounce price afterward. For example, in the post-1934 pre-1968 era, an assay stated as $3.50 per ton was equivalent to 0.1 ounce of gold per ton.

hanging wall. The wall or rock above an inclined ore body; the upper side of the ore in relation to the dip of the ore body. *Contrast* **footwall.**

hydrothermal processes. The "hot water" processes in which heated, dilute, water-rich solutions dissolve, transport, and precipitate the metallic and associated nonmetallic minerals that form most lode mineral deposits. In some cases, the hot solutions are derived directly from cooling igneous magmas; in other cases, they are fossil surface waters that were heated as their host rocks were buried. Water is the chief component of the solutions, but water by itself does not dissolve significant amounts of heavy metals. Additional components, including chlorine and sulfur, form stable complex ions with heavy metals. Metals are transported as water-soluble complex ions and are deposited when the complex ions break down.

hypogene deposit. A mineral deposit formed at or in the crust of the earth from ascending hydrothermal solutions; sometimes called *primary deposits.* These include *syngenetic deposits* formed on the sea floor by sedimentary and volcanic hydrothermal processes, as well as *epigenetic deposits. Contrast* **supergene deposit.**

leached cap or capping. The weathered zone overlying a sulfide deposit. The term *leached cap* is generally used for the zone overlying a disseminated deposit, i.e., a porphyry copper deposit. The term *gossan* is used for the oxidized zone that overlies veins and massive deposits. A leached cap or gossan consists mainly of oxidized minerals, primarily hydrated iron oxides and sulfates; its mineralogy and texture depend on the nature of the underlying sulfide deposit. The nonhydrated iron oxide, hematite, is a common component of gossan that overlies rich deposits of chalcocite. Relatively insoluble native metals—gold, silver, and copper—may be concentrated in some leached capping.

level. *See under* **mine workings.**

lode. In a limited sense, a vein; generally a deposit of metallic minerals contained within a bedrock host. The lode may have sharp or gradational boundaries with its host.

metasomatism. Essentially the simultaneous solution of an existing mineral and deposition of a new mineral at the capillary scale *(replacement),* generally at constant volume.

mine workings. Openings or excavations for extracting valuable minerals. They are divided generally into underground and open-cut, or surface, workings.

An *adit* is a nearly horizontal passage driven from the surface for the working or unwatering of a mine. Many main haulage or drainage adits are called *tunnels,* a term that is also applied to a nearly horizontal passage with two openings.

A *crosscut* is a nearly horizontal passage driven across the rock or vein structure, usually for access or exploration. *Contrast drift.*

A *decline* (or *incline*) is a passage, usually driven by trackless methods, that has a substantial slope. For haulage purposes, declines or inclines are generally at a grade of about 15 percent or less.

A *drift* is an underground passage that follows a vein or other tabular ore body. *Contrast crosscut.*

A *level* is a main, nearly horizontal, underground passage or roadway.

A *raise* is a vertical or steeply inclined passage driven upward from a level. *Contrast winze.*

A *shaft* is a vertical or steeply inclined passage driven from the surface or underground workings. It is fitted with machinery for hoisting ore, waste, and men from the mine workings.

A *stope* is an opening in a mine made for the purpose of extracting ore. The outlines of the stope approximate the outlines of the ore. A *blast-hole stope* is generally a variant of an open stope using long drill holes. A *cut-and-fill stope* enables mining in upward or lateral slices. After the ore is blasted and removed, the opening is filled with waste. It is a versatile method that may be used with relatively weak rocks. An *open stope* allows the removal of ore in strong wall rocks without the use of artificial support, except for local pillars or cribs. The Lucky Shot mine was generally open stoped. A *shrinkage stope* is an opening mined in successive upward slices. Miners stand on broken ore and drill upward; after the next blast, ore is withdrawn from chutes at the bottom of the stope, leaving enough for the miners to stand on and drill the next upward slice. The broken ore left in the stopes helps to support the walls of the stope. Shrinkage stoping is used in steep veins with moderate to strong walls. The shrinkage stopes at the Golden Zone mine were as much as twenty-five feet across. A *glory hole* is a funnel-shaped opening at the surface that connects, via one or more raises, to an underground haulage level. Ore is broken by drilling benches around the periphery of the glory hole; the pieces then fall to the underlying raise or mill hole. A combination of stopes is commonly used in each mine. Glory-hole mining is relatively inexpensive. Dunkle proposed to glory-hole the Golden Zone ore body. Because he could not complete the haulage adit below the proposed glory hole, he mined the ore by shrinkage stoping in 1941.

A *winze* is an opening driven downward from a level. *Contrast raise.*

ore. A mineral or mineral aggregate containing precious or useful metals or metalloids that are present in such quantity, grade, and chemical combination as to make commercial extraction possible. To allow for political influences, Cameron (*At the Crossroads,* 7, see also 8–20) defines ore as "material from which one or more useful minerals or metals can be extracted under current economic and technologic conditions in a politically and socially acceptable manner."

ore shoot. An area of payable ground (ore) included within a mass of lower-grade rock.

placer. Most commonly a water-borne deposit of sand and gravel that contains particles of dense metallic minerals, such as gold, platinum, and magnetite, which were eroded from an original bedrock host and have been concentrated by natural density sorting. In most alluvial (riverine) deposits, the dense minerals are concentrated on the bedrock surface at the base of the column of sand and gravel.

porphyry. An igneous rock with crystals of varying size. *Porphyry* is often used to describe any fine-grained, mineralized granitic rock, such as *porphyry copper.*

raise. *See under* **mine workings.**

replacement. *See* **metasomatism.**

reserve. "*Identified resources* of mineral- or fuel-bearing rock from which the mineral or fuel can be extracted profitably with existing technology and under present economic conditions" (Bates and Jackson, *Glossary of Geology,* 531). Reserves are classified along probability lines based on the continuity of the ore body and how much work has been done to define it. The common categories, from most probable downward, are *proven, probable,* and *possible.* The U.S. Bureau of Mines used *measured, indicated,* and *inferred,* respectively, for approximately the same categories.

shaft. *See under* **mine workings.**

sill. A tabular igneous intrusion emplaced on a structural plane that is parallel to the bedding or layering of the rock. *Contrast* **dike.**

skarn (or tactite). A coarse-grained rock composed of garnet, epidote, iron-rich pyroxene, wollastonite, and less commonly, scapolite formed at or near the contact between an igneous intrusive rock and a calcic rock, commonly limestone or dolomite. Locally it contains native gold as well as copper, molybdenum, tungsten, and zinc, which, respectively, are most commonly formed as chalcopyrite, molybdenite, scheelite, and sphalerite. In 1923 and 1924, Dunkle explored zinc-rich skarn deposits in the Copper Mountain, or Mt. Eielson, area of Mt. McKinley National Park.

stope. *See under* **mine workings.**

strategic mineral. A mineral that is essential to the national defense of a country, which is wholly or largely dependent on external sources for that mineral. A strategic ore may be mined at a loss during a national emergency.

strike. "The direction or trend taken by a structural surface, e.g., a bedding or fault plane, as it intersects the horizontal" (Bates and Jackson, *Glossary of Geology,* 618). *See also* **dip.**

supergene deposit. A mineral deposit formed from descending waters of surface origin; sometimes called *secondary deposits.* In sulfide-bearing mineral deposits exposed at the surface, acidic waters, formed by the oxidation of sulfide minerals, leach soluble metals above the water table and redeposit them below the water table. These are essentially the same reactions as in acid-rock drainage (often abbreviated ARD), the natural counterpart of acid-mine drainage (so-called AMD). In copper-bearing mineral deposits, chalcocite replaces hypogene chalcopyrite or bornite below the water table, forming an enriched supergene deposit. Gold, which is relatively insoluble, can become enriched and concentrated above the water table because of the leaching of the more soluble minerals. Because of Alaska's present and geologically recent cold climate, significant supergene deposits of recent origin are rare. Dunkle, however, was very familiar with supergene deposits and processes from his work at Ely and Contact, Nevada. *Contrast* **hypogene deposit.**

syncline. A fold of stratified rocks whose inner layers contain the younger rocks; it is generally concave upward. *Contrast* **anticline.**

syngenetic deposit. A mineral deposit formed at the same time as its host rock. The deposit at the Beatson mine on Latouche Island is probably syngenetic. *See also* **diagenetic deposit** and **epigenetic deposit.**

tactite. *See* **skarn.**

washing plant. A mining machine that combines two or more of the necessary functions of a placer-mining plant: classifying (sizing) the gold-bearing gravel, recovering the gold, and disposing of the tailings. The washing plant designed by Washington Iron Works of Seattle and used by Dunkle at the Caribou Creek property classified the gravel with a rotating screen *(trommel),* recovered the gold in short sluice boxes, and stacked the tailings with a conveyor belt. It was mobile and was fed by a dragline. Bulldozers were used to strip the overburden and to clean the floor of the pit.

winze. *See under* **mine workings.**

SELECTED BIBLIOGRAPHY FOR GLOSSARY

Bates, Robert L., and Julia A. Jackson. *Glossary of Geology.* 2nd ed. Falls Church, Va.: American Geological Institute, 1980.

Cameron, E. N. *At the Crossroads: The Mineral Problems of the United States.* New York: John Wiley & Sons, 1986.

Lindgren, Waldemar. *Mineral Deposits.* 4th ed. New York: McGraw-Hill, 1933.

Park, Charles F., Jr., and R. A. MacDiarmid. *Ore Deposits.* 2nd ed. San Francisco: W. H. Freeman, 1970.

Thrush, Paul, comp. *Dictionary of Mining Terms.* Washington, D.C.: U.S. Bureau of Mines, 1968; Chicago: McLean Hunter, 1990.

U.S. Department of Agriculture, Forest Service. *Anatomy of a Mine: From Prospect to Production.* General Technical Report INT-35. Ogden, Utah: Intermountain Forest and Range Experiment Station, 1977.

Bibliography

Some citations include abbreviations from the list at the beginning of the book.

INTERVIEWS

All interviews were conducted by the author unless otherwise noted.

Andresen, Decema Kimball. Anchorage, 18 June 1997.

Atwood, Robert B. Anchorage, 15 and 20 February 1996.

Bagoy, Peter Sr., with Pearse Walsh. Anchorage, 19 and 23 April 1996.

Dunkle, Bruce B. Anchorage, 26 June 1996.

Dunkle family (Mr. and Mrs. Bruce B. Dunkle, William E. Dunkle, and Mr. and Mrs. John H. Dunkle), with Harry and Gloria Bowman. Various locales in southern California, 29–31 March and 1 April 1997.

Ellis, Paul. Palmer, Alaska, 20 and 27 April 1996.

Miscovich, John. Anchorage, 22 July 1993.

Neubauer, Jack. Fairbanks, March 1996.

Noey, Luther. Interview by telephone; Mr. Noey was in Tyler, Tex., 7 April 1996. The interview was followed by correspondence.

Stanford, Charlotte Williams (Mrs. James Stanford). Notes from interviews with Barbara Bell, ca. 1980. Notes furnished by David Kunzmann, Burdett, N.Y., 1996.

Till, Anna Marie (Mrs. Leo J. Till) and Vincent Harrold. Mt. Angel, Oreg., 23 June 1996.

Waugaman, William A. Fairbanks, October 1996.

UNPUBLISHED REPORTS

Accolade Mines, Inc., mainly by Billie Dunkle. "Consolidated Report, 1961." Miscellaneous Report MR-67-9, ADGGS, 1961.

Apell, G. A. "Broad Pass Coal Reports, Week of August 8–12, 1944." Miscellaneous Report MR-67-4, ADGGS, 1944.

Bigelow, C. A. "Evaluation Report on Caribou Creek." Private engineer's report for Hammon Consolidated Goldfields, San Francisco, Calif., 1925.

Buzzell, Rolfe G. "Caribou Creek Drainage History." Excerpt from "Drainage Histories of the Kantishna Mining District, 1903–1968," 13 p. Typed manuscript, National Park Service, Anchorage, 1989.

———, ed. "Flat and Iditarod, 1993–1995 Oral History Interviews." Open-File Report 66, U.S. Bureau of Land Management, Anchorage, 1997.

Buzzell, Rolfe G., and Darrell L. Lewis. "Historic Building Survey Report, Flat, Alaska." Open-File Report 64, U.S. Bureau of Land Management, Anchorage, 1997.

Carlisle, H. C. "Early Days of Kennecott, Alaska." Typed interview of L. A. Levensaler, [ca. 1960]. APRD.

Douglass, W. C. "A History of the Kennecott Mines, Kennecott, Alaska." Miscellaneous Report MP-21, ADGGS, 1964, reprint 1971.

Dunkle, Peter S. "Thirty-Six Feet of Boys [ca. 1926]." Excerpt from " 'Dunkle'-Dusk-Twilight-Evening," by Clara Adella Dresskell Page, 18 p. Typed report, 1977. DF.

———. "Tradition and Times [ca. 1926]." Excerpt from " 'Dunkle'-Dusk-Twilight-Evening," by Clara Adella Dresskell Page, 9 p. Typed report, 1977. DF.

Dunkle, Wesley Earl (WED). "Class Notes on Mineralogy and Crystallography." Sheffield Scientific School, Yale University, New Haven, 1907. Memorabilia, collection of Melvin and Mary J. Barry, Anchorage.

———. "File No. 78, Nikolai Mine." Letter report to Stephen Birch, 6 December 1915, 1 p. and map. Author's files.

———. "Mining Properties Visited in Alaska in 1924." Letter report to E. T. Stannard, 4 December 1924, 11 p. In file "Healy Quadrangle, Historic 1920–1945." Kennecott Exploration Co., Salt Lake City.

———. "The Golden Zone Mine: Report to Shareholders." Miscellaneous Report MR-67-6, ADGGS, 1938.

———. "Golden Zone Mine, Inc." Report available with 1938 report (above) as Miscellaneous Report MR-67-6, ADGGS, 1947.

———. "Economic Geology and History of the Copper River District." Miscellaneous Report MR-87-4, ADGGS, 1954.

Erickson, Arthur F. "Memorandum of Caribou Creek in Kantishna District." Manuscript, Carrington and Co., Anchorage, 1938. Author's files.

Hawley, C. C., and Associates, Inc. "Mineral Appraisal of Lands Adjacent to Mt. McKinley National Park." Open-File Report 24-78, U.S. Bureau of Mines, Juneau, 1978.

Holzheimer, F. W. "Lode Mining Activity, Otter Creek, Iditarod District." Miscellaneous Report MR-73-1, with Roehm report, this section, ADGGS, ca. 1926.

Keller, H. A. "Mother Lode Copper Mines Company Report." 1 August 1911. Packet C, File 140, Document 1. Kennecott Copper Corp., Salt Lake City.

Kennecott Copper Corp. "Annual Report for 1938, Alaska Mines, Kennecott Copper Corp." Typed copy furnished by Ronald N. Simpson, Copper Center, Alaska, 1938.

March, James. "Report of Coal Drying at Colorado Station, November–December 1957, Using the Dunkle Prototype Dryer." Private report, Earl H. Beistline collection, Fairbanks, January 1958.

McKay, Ralph. "Appertaining to the History of the Copper River Area, Alaska." Excerpts from L. A. Levensaler letters. APRD, 1955–1967.

Moerlein, George. "Golden Zone Property Evaluation." Typed report. Bear Creek Mining Co. files at Kennecott Exploration Co., Spokane, Wash., 1961.

Page, Clara Adella Dresskell. " 'Dunkle'-Dusk-Twilight-Evening." Typed genealogical report, 1977. DF.

Potter, Ocha. "Alaska." Typed report. Warren McCullough collection, Helena, Mont., ca. 1940.

Richelson, W. A. "List of Reports on Prospects Examined, Alaska Development and Mineral Co." Typed report. Kennecott Exploration Co., Salt Lake City, 1945.

Roehm, J. C. "Preliminary Report of Golden Horn Mine, Iditarod Mining District, Alaska." Miscellaneous Report PE-73-1, with Holzheimer report, this section, ADGGS, 1926.

Shedwick, William John, Jr. Untitled typed journal. Peggy Shedwick McCook collection, Dallas, Tex., ca. 1965.

Smith, Alexander. "Report on the Golden Zone Mine." Typed report. Ridgeway R. Wilson & Assocs., Consulting Engineers, Vancouver, Canada, ca. 1945. Author's files.

Till, Anna Marie. "Memories of Anna Marie Till's Life." Typed autobiography, Mt. Angel, Oreg., 1993. Author's files.

NEWSPAPERS

Bylined articles in newspapers are listed under "Published Sources" in the bibliography. All others are cited only in the notes section.

Anchorage Daily News
Anchorage Daily Times
Anchorage Sunday Times, 1975
Cook Inlet Pioneer (Anchorage), 1915, succeeded *Knik News*
Cordova Daily Alaskan
Cordova Daily Times
Fairbanks Daily News-Miner
Jessen's Weekly (Fairbanks), 1958
Knik News, 1915
Lewiston (Idaho) *Tribune,* 1921
Nome Nugget, 1920
Oberlin (Ohio) *Record,* 1914
Pittsburgh Post-Gazette, 1931
Sandusky (Ohio) *Daily Register,* 1905
Sandusky (Ohio) *Star Journal,* 1905
The Scotsman (Edinburgh, Scotland), 1955
Seattle Post-Intelligencer

PUBLISHED SOURCES

Agricola, Georgius. *De Re Metallica* (1556). Translated by Herbert C. and Lou Henry Hoover. London: The Mining Magazine, 1912.

Allen, Lt. Henry T. "An Expedition to the Copper, Tanana, and Koyukuk Rivers of Alaska in 1885." *The Alaska Journal* 15, no. 2 (1985): 13–93. Reprint of *Report on the Expedition to the Copper, Tanana, and Koyukuk Rivers in the Territory of Alaska in the Year 1885.* Washington, D.C.: GPO, 1887.

Anchorage Centennial Commission, Aviation Committee. *Honoring 100 Alaska Bush Pilots: Alaska Purchase Centennial, 1867–1967.* Anchorage: Anchorage Centennial Commission, 1967.

Andresen, Decema Kimball. *Memories of Latouche.* Anchorage: Publication Consultants, Inc., 1997.

Atwood, Evangeline. *Anchorage: All-American City.* Portland, Oreg.: Binfords & Mort, 1957.

———. *Frontier Politics: Alaska's James Wickersham.* Portland, Oreg.: Binford & Mort, 1979.

Atwood, Robert B. "Between Us." *Anchorage Sunday Times,* 19 January 1975, A-5.

Atwood, W. W. *Geology and Mineral Resources of Parts of the Alaska Peninsula.* USGS Bulletin 467. Washington, D.C.: GPO, 1911.

Baragwanath, John [John Gordon]. *Pay Streak.* New York: Doubleday Doran, 1936.

———. *A Good Time Was Had.* New York: Appleton-Century-Crofts, 1962.

Barry, Mary J. *Seward Alaska: A History of the Gateway City.* Vol. 2, *1914–1923, The Railroad Construction Years.* Anchorage: MJP Barry, 1993.

Bartlett, Bob. "Dimond Brings Alaska's Air Needs to Fore." *FDNM,* 12 April 1934.

Bateman, Alan M., and D. H. McLaughlin. "Geology of the Ore Deposits of Kennecott, Alaska." *Economic Geology* 15, no. 1 (1920): 1–80.

Beach, Rex E. *The Iron Trail: An Alaskan Romance.* New York: A. L. Burt, 1913.

Belin, G. d'Andelot. *Thirty-Year History, Class of 1908, Sheffield Scientific School, Yale University.* New Haven: Yale University Press, 1940.

Bennett, Russell H. *Quest for Ore.* Minneapolis: T. S. Denison & Co., 1963.

Berg, Rhinehard. "Rhinehard Berg: A Twentieth-Century Pioneer." *The Alaska Miner* 26 (1998): pt. 1, no. 3, 10–13, 18; pt. 2, no. 4, 10–12; pt. 3, no. 5, 10–11, 14.

Bernstein, Peter L. *The Power of Gold: The History of an Obsession.* New York: John Wiley, 2000.

Bleakley, Geoffrey T. *A History of the Chisana Mining District, Alaska, 1890–1990.* National Park Service, Resources Report NPS/AFARCR/CRR-96/29, 1996.

Book of Biographies, 37th Judicial District. Buffalo, N.Y.: Biographical Publishing Co., 1899.

Bordman, Gerald. *American Operetta from H. M. S. Pinafore to Sweeney Todd.* New York: Oxford University Press, 1981.

Borthwick, John David. *Three (3) Years in California.* Edinburgh, Scotland: William Blackford and Sons, 1857; reprint, Oakland, Calif.: Biobooks, 1948.

Bowen, Ezra, ed. *This Fabulous Century, 1900–1910.* Vol. 1. New York: Time-Life Books, 1969.

Brandt, Nat. *The Town That Started the Civil War.* Syracuse: Syracuse University Press, 1990.

Brown, Robert L. *Central City and Gilpin County, Then and Now.* Caldwell, Idaho: Caxton Printers, 1994.

Browne, Belmore. *The Conquest of Mount McKinley: The Story of Three Expeditions Through the Alaska Wilderness to Mount McKinley.* New York: G. P. Putnam's Sons, 1913.

Buhro, Harry. *Rough-Stuff and Moonlight: Of Deeper Love Hath None.* Philadelphia: Dorrance, 1948.

Bunyak, Dawn. "To Float or to Sink: A Brief History of Flotation Milling." In *Mining History Association 2000 Journal,* 7th annual, 35–44. Denver: Mining History Association, 2000.

Buzzell, Rolfe G. Introduction to *Memories of Old Sunrise: Gold Mining on Alaska's Turnagain Arm. Autobiography of Albert Weldon Morgan,* edited by Rolfe G. Buzzell, xi–xxiii. Anchorage: Cook Inlet Historical Society, 1994.

Capps, S. R. *The Willow Creek District, Alaska.* USGS Bulletin 607. Washington, D.C.: GPO, 1915.

———. "Gold Lode Mining in the Willow Creek District." In USGS Bulletin 692, 177–186. Washington, D.C.: GPO, 1919.

———. "Mineral Resources of the Western Talkeetna Mountains." In USGS Bulletin 692, 187–205. Washington, D.C.: GPO, 1919.

———. "Mineral Resources of the Upper Chulitna Region." In USGS Bulletin 692, 207–232. Washington, D.C.: GPO, 1919.

Carlisle, H. C. "David D. Irwin, an Interview." *Mining Engineering* (June 1964): 88–91.

Carlson, Phyllis D. "Alaska's Hall of Fame Painter: A Sourdough Painter Preacher." *Alaskana* 1, no. 10 (1971): 1–3.

Carter, M. [Marilyn]. *The Iditarod Trail: The Old and the New.* Palmer, Alaska: Aladdin Publishing, 1990.

Chittenden, Russell H. *History of the Sheffield Scientific School of Yale University, 1846–1922.* 2 vols. New Haven: Yale University Press, 1928.

Church, Howard E., ed. *Class History, 1908, Sheffield Scientific School.* Vol. 1. New Haven: Yale University Press, 1908.

CIM (Canadian Institute of Mining and Metallurgy). "A. Smith" (obituary). In "Members in the Spotlight." *CIM Bulletin* 89 (May 1996): 16.

Clifton, Howard. *Rails North.* Seattle: Superior Publishing Co., 1981.

Cole, Dermot. *Fairbanks: A Gold Rush Town That Beat the Odds.* Fairbanks: Epicenter Press, 1999.

Cole, Terrence. "A History of the Nome Gold Rush: The Poor Man's Paradise." Ph.D. diss., University of Washington, 1983.

———. *Crooked Past: The History of a Frontier Mining Camp, Fairbanks, Alaska.* Fairbanks: University of Alaska Press, 1991.

Collier, A. J., F. L. Hess, P. S. Smith, and A. H. Brooks. *The Gold Placers of Parts of the Seward Peninsula, Alaska.* USGS Bulletin 328. Washington, D.C.: GPO, 1908.

Committee on the Territories. *Railroads in Alaska.* U.S. House of Representatives, 60th Cong., 1st sess. Washington, D.C.: GPO, 1908.

Coward, Noel. *Present Indicative.* New York: Doubleday, Doran & Co., 1937.

———. *Future Indefinite.* Garden City, N.Y.: Doubleday & Co., 1954.

Craig, J. R. "The Cu-S System." In "Sulfide Mineralogy," chap. 5, in *Sulfide Phase Equilibria,* CS58–CS63. Vol. 1, Sulfide Mineralogy Short Course Notes. Mineralogical Society of America, November 1974.

Crittenden, Katherine Carson. *Get Mears! Frederick Mears: Builder of the Alaska Railroad.* Portland, Oreg.: Binford and Mort, 2002. Published in cooperation with Cook Inlet Historical Society, Anchorage.

Cronin, Vincent. *Paris on the Eve, 1900–1914.* New York: St. Martin's Press, 1991.

Cronon, William. "Kennecott Journey: The Paths Out of Town." In *Under an Open Sky: Rethinking America's Western Past,* edited by William Cronon, George Miles, and Jay Gitlin, 28–51, 276–281. New York: W. W. Norton, 1992.

Dall, William H. *Alaska and Its Resources.* Boston: Lee and Shepard, 1870; reprint, New York: Arno and New York Times, 1970.

Davis, Richard Harding. *Soldiers of Fortune.* New York: Scribners, 1897.

DiMarchi, Jack, and Bob Friesen. "An Update on the Recent Underground Advanced Exploration Project at Pogo." In *Abstracts, 2000 Annual Convention.* Anchorage: Alaska Miners Association, 2000.

Dixon, Colin J. *Atlas of Economic Mineral Deposits.* Ithaca, N.Y.: Cornell University Press, 1979.

———. "The Luanshya Copper Deposit, Zambia." In *Atlas of Economic Mineral Deposits,* 42–43. Ithaca, N.Y.: Cornell University Press, 1979.

———. "The Platinum Deposits of the Merensky Reef, South Africa." In *Atlas of Economic Mineral Deposits,* 102–103. Ithaca, N.Y.: Cornell University Press, 1979.

Duggan, E. J. "Ammonia Leaching at Kennecott." AIME *Transactions* 106 (1933): 547–588.

Dunkle, Peter S. "On Oil History and Discovery." *Lewiston (Idaho) Tribune,* 18 September 1921.

Dunkle, William E. *A Birds Eye View.* Camarillo, Calif.: Juli Sellers Press, 1988.

Emmons, S. F. "The Secondary Enrichment of Ore Deposits." AIME *Transactions* 30 (1901): 177–217.

Emmons, W. H. "Recent Progress in Studies of Supergene Enrichment." In *Ore Deposits of the Western States,* 386–418. New York: AIME, 1933.

Fetherling, Douglas. *The Gold Crusades: A Social History of Gold Rushes, 1849–1929.* Toronto: Macmillan of Canada, 1988.

Fletcher, Robert S. *A History of Oberlin College: From Its Foundation Through the Civil War.* 2 vols. Oberlin, Ohio: Oberlin College, 1943.

Foote, Mary Hallock. *The Led-Horse Claim: A Romance of a Mining Camp.* New York: James R. Osgood, 1883.

———. *A Victorian Gentlewoman in the Far West: The Reminiscences of Mary Hallock Foote.* Edited and with an introduction by Rodman W. Paul. San Marino, Calif.: Huntington Library, 1972.

Ford, James L. "Helen Green's *At the Actor's Boarding House.*" Book review. *The Bookman* 25 (1907): 431–433.

Gille, F. A., ed. *The Encyclopedia of Pennsylvania.* New York: Somerset, 1983.

Grant, U. S., and D. F. Higgins Jr. "Copper Mining and Prospecting on Prince William Sound." In USGS Bulletin 379, 87–96. Washington, D.C.: GPO, 1909.

Graton, L. C., and J. Murdoch. "The Sulphide Ores of Copper." AIME *Transactions* 45 (1914): n.p.

Graumann, Melody Webb. *Big Business in Alaska: The Kennecott Mines, 1898–1938.* National Park Service, Cooperative Park Studies Unit, Occasional Paper No. 1. Denver: Technical Information Center, 1977.

Green, Lewis. *The Gold Hustlers.* Anchorage: Alaska Northwest Publishing Co., 1977.

Hammond, John Hays. *The Autobiography of John Hays Hammond.* 2 vols. New York: Farrar & Rinehart, 1935.

Harris, Arland S. Introduction and annotation to *Schwatka's Last Search: The New York Ledger Expedition Through Unknown Alaska and British America,* including "The Journal of Charles Willard Hayes, 1891." Fairbanks: University of Alaska Press, 1996.

Hawley, Charles Caldwell. "Wesley Earl Dunkle: The Years for Stephen Birch, 1910–1929." *Mining History Journal,* 5th annual (1998): 89–102.

Hawley, C. C., and Allen L. Clark. *The Geology and Mineral Deposits of the Upper Chulitna District, Alaska.* USGS Professional Paper 758-B. Washington, D.C.: GPO, 1974.

"'Helen Green' in Chronicle and Comment." *The Bookman* 27 (March 1908): 3–7.

"'Helen Green Van Campen' in Chronicle and Comment." *The Bookman* 37 (1913): 364–365.

Hellenthal, J. A. *The Alaskan Melodrama.* New York: Liveright Publishing, 1936.

Holbrook, Stewart H. *The Age of the Moguls.* Mainstream of America Series, edited by Lewis Gannett. New York: Doubleday, 1953.

Hoover, Herbert C. *The Memoirs of Herbert Hoover: The Years of Adventure, 1874–1920.* New York: Macmillan, 1951.

Hoover, Herbert C., and Lou Henry Hoover, trans. *De Re Metallica* (1556), by Georgius Agricola. London: The Mining Magazine, 1912.

Ickes, Harold L. *Not Guilty. An Official Inquiry Into the Charges Made by Glavis and Pinchot Against Richard A. Ballinger, Secretary of the Interior, 1909–1911.* Washington, D.C.: GPO, 1940.

Janson, Lone E. *The Copper Spike.* 2nd ed. Anchorage: HaHa, Inc., 1975.

Johnson, B. L. "Mining on Prince William Sound." In Bulletin 692, 143–151. Washington, D.C.: GPO, 1919.

———. "Mineral Resources of the Jack Bay District and Vicinity, Prince William Sound." In Bulletin 692, 153–173. Washington, D.C.: GPO, 1919.

Jones, David L., N. J. Silberling, and David Newhouse. "Wrangellia—A Displaced Terrane in Northwestern North America." *Canadian Journal of Earth Science* 14 (1978): 2365–2377.

Jones, Stan. "Another Spin on 'Wizard of Oz.' " *Anchorage Daily News,* 22 September 1996.

———. "Unlikely Savior." *Anchorage Daily News,* 22 September 1996.

Joralemon, Ira B. *Copper: The Encompassing Story of Mankind's First Metal.* Berkeley: Howell-North Books, 1973.

———. *Adventure Beacons.* Edited by Peter Joralemon. New York: Society of Mining Engineers of AIME, 1976.

Kelley, Brooks Mather. *Yale: A History.* New Haven: Yale University Press, 1974.

Kennedy, Kay J. "Forty Men Working at Golden Zone." *FDNM,* 7 July 1939.

———. "Helen Van Campen Dies Here." *FDNM,* 15 August 1960.

Koschmann, A. H., and M. H. Bergendahl. *Principal Gold-Producing Districts of the United States.* USGS Professional Paper 610. Washington, D.C.: GPO, 1968.

Laguna, Frederica de. *Chugach Prehistory: The Archaeology of Prince William Sound, Alaska.* Seattle: University of Washington Press, 1956, reprint 1967.

LaPointe, Daphne D., Joseph V. Tingley, and Richard B. Jones. *Mineral Resources of Elko County, Nevada.* Bulletin 106. Reno: Nevada Bureau of Mines and Geology, 1991.

Le Conte, Joseph. *Elements of Geology: A Textbook for Colleges and for the General Reader.* 5th ed. Revised by Herman Le Roy Fairchild. New York: D. Appleton and Co., 1907.

Leonard, John W., ed. *Women's Who's Who in America.* New York: American Commonwealth, 1915.

Lethcoe, Jim, and Nancy Lethcoe. *A History of Prince William Sound, Alaska.* Valdez, Alaska: Prince William Sound Books, 1994.

———. *Valdez Gold Rush Trails of 1898–99.* Valdez, Alaska: Prince William Sound Books, 1996.

Lincoln, Francis Church. "The Big Bonanza Copper Mine of Latouche Island, Alaska." *Economic Geology* 4, no. 3 (1909): 201–213.

Lindgren, Waldemar. "Metasomatic Processes in Fissure Veins." AIME *Transactions* 30 (1901): 578–692.

———. "The Character and Genesis of Certain Contact Deposits." AIME *Transactions* 31 (1902): 226–244.

———. "The Relationship of Ore-Deposition to Physical Conditions." *Economic Geology* 2 (1907): 105–127.

———. *Mineral Deposits.* 4th ed. New York: McGraw-Hill Book Co., 1933.

Linforth, F. A. "Application of Geology to Mining in the Ore Deposits of Butte, Montana." In *Ore Deposits of the Western States,* 695–701. New York: AIME, 1933.

Lord, Walter. *The Good Years: From 1900 to the First World War.* New York: Harper, 1960.

Loughlin, G. F., and C. H. Behre Jr. "Classification of Ore Deposits." In *Ore Deposits of the Western States,* 17–55. New York: AIME, 1933.

MacKevett, E. M., Jr., and others. "Kennecott-Type Deposits in the Wrangell Mountains, Alaska: High-Grade Copper Ores Near a Basalt-Limestone Contact." In *Mineral Deposits of Alaska,* 66–89. Economic Geology Monograph 9. N.p.: Economic Geology Publishing Co., 1997.

MacLean, Robert Merrill, and Sean Rossiter. *Flying Cold: The Adventures of Russel Merrill, Pioneer Alaskan Aviator.* Fairbanks: Epicenter Press, 1994.

Maddren, A. G. "Gold Placer Mining Developments in the Innoko-Iditarod Region, Alaska." In USGS Bulletin 480-I, 236–270. Washington, D.C.: GPO, 1911.

Martin, G. C. *Geology and Mineral Resources of the Controller Bay Region, Alaska.* USGS Bulletin 335. Washington, D.C.: GPO, 1908.

McDonald, Lucile. "Alaska Steam: A Pictorial History of the Alaska Steamship Company." *Alaska Geographic* 11, no. 4 (1984): n.p.

McLaughlin, D. H., and Reno H. Sales. "Utilization of Geology by Mining Companies." In *Ore Deposits of the Western States,* 683–694. New York: AIME, 1933.

Mendenhall, W. C., and F. C. Schrader. *The Mineral Resources of the Mt. Wrangell District, Alaska.* USGS Professional Paper 15. Washington, D.C.: GPO, 1903.

Mills, Stephen E., Jr., and James W. Phillips. *Sourdough Sky: A Pictorial History of Flights and Flyers in the Bush Country.* Seattle: Superior Publishing Co., 1969.

"Mining Engineer." Review of *Pay Streak,* by John Baragwanath. *Time* 28 (30 November 1936): 73–75.

Morrell, W. P. *The Gold Rushes.* Pioneer Histories, edited by V. T. Harlow and J. A. Williamson. Chester Springs, Pa.: Dufour, 1968.

Morris, Richard B., ed. *Encyclopedia of American History.* New York: Harper & Brothers, 1953.

Naske, Claus-M., and Herman E. Slotnick. *Alaska: A History of the 49th State.* Grand Rapids, Mich.: William B. Eerdmans, 1979.

Navin, Thomas R. *Copper Mining and Management.* Tucson: University of Arizona Press, 1978.

Nichols, Jeannette Paddock. *Alaska: A History of Its Administration, Exploitation, and Industrial Development During Its First Half Century Under the Rule of the United States.* Cleveland: Arthur H. Clark Co., 1924.

Oberlin College. "Florence Hull." In *Hi-O-Hi,* n.p. Student annual. Oberlin, Ohio: Oberlin College, 1908.

O'Connor, Harvey. *The Guggenheims: The Making of an American Dynasty.* New York: Covici, Friede, 1937; reprint, New York: Arno, 1976.

Oliver, Howard T. Letter quoted in "Wesley Earl Dunkle—1908 Class Notes," compiled by Starr Barnum. *Yale Alumni Magazine* (December 1957): n.p.

Oliver, Simeon (Nutchuk), with Alden Hatch. *Son of the Smoky Sea.* New York: Julian Messner, 1941.

Paige, Sydney, and Adolph Knopf. "Reconnaissance in the Matanuska and Talkeetna Basins, Alaska, With Notes on Placers of the Adjacent Region." In USGS Bulletin 314-F, 104–125. Washington, D.C.: GPO, 1907.

Paul, Rodman W., ed. *A Victorian Gentlewoman in the Far West: The Reminiscences of Mary Hallock Foote,* by Mary Hallock Foote. San Marino, Calif.: Huntington Library, 1972.

Peeke, H. L. "Bar Honors Dead Judge." *Sandusky Daily Register,* 6 June 1906, 1, 3.

Penrose, R. A. F., Jr. "The Superficial Alteration of Ore Deposits." *Journal of Geology* 2 (1894): 288–317.

Pioneers of Alaska, Igloo 15, Auxiliary 4. *Fond Memories of Anchorage Pioneers.* Vol. 1. Anchorage: Pioneers of Alaska, 1996.

Porter, Jean. *The Flying North.* New York: Macmillan, 1965.

Posnjak, E., E. T. Allen, and H. E. Merwin. "Some Reactions Involved in Secondary Enrichment." Contribution 7, Secondary Enrichment Investigation. *Economic Geology* 10 (1915): 491–535.

Quinn, A. O. *Iron Rails to Alaskan Copper: The Epic Triumph of Erastus Corning Hawkins,* 2nd ed. Quaker Mountain, Whiteface, N.Y.: D'Aloquin Publishing Co., 1995.

Ransome, F. L. "Historical Review of Geology as Related to Western Mining." In *Ore Deposits of the Western States,* 1–16. New York: AIME, 1933.

Ray, J. C. "The Willow Creek Gold-Lode District, Alaska." In USGS Bulletin 849-C, 165–229. Washington, D.C.: GPO, 1933.

Ray, Marie Beynon. "Throwing a Party." *Colliers* 90 (31 December 1932): 24, 41.

Ray, R. G. *Geology and Ore Deposits of the Willow Creek Mining District, Alaska.* USGS Bulletin 1004. Washington, D.C.: GPO, 1954.

Redding, Robert H. *North to the Wilderness: The Story of an Alaskan Boy.* New York: Doubleday, 1970.

Reiss, Marguerite. "The Alaska Miner of West Point." *The Alaska Miner* 22 (October 1994): 9, 13–14.

Renkert, Robert, and others, comps. "Weimer, Robert E. (Pete) and Judy." In *Fond Memories of Anchorage Pioneers,* vol. 1, 38–40. Anchorage: Pioneers of Alaska, Igloo 15, Auxiliary 4, 1996.

Review of *All That Glitters,* in "New Plays in Manhattan." *Time* 30 (31 January 1938): 39–40.

Ricci, Inger Jensen. "Childhood Memories of Kennecott." In "Wrangell–St. Elias International Mountains Wilderness Area." *Alaska Geographic* 8, no. 1 (1981): 80–89.

Roderick, Jack. *Crude Dreams: A Personal History of Oil and Politics in Alaska.* Fairbanks: Epicenter Press, 1997.

Roppel, Patricia. " 'Have I Got a Deal for You.' Mining Frauds on Douglas Island." *Alaska History* 5 (1990): 18–22.

Rosenthal, Eric. *Gold! Gold! Gold! The Johannesburg Gold Rush.* London: Macmillan, 1970.

Ross, Clyde P. "Mineral Deposits Near the West Fork of the Chulitna River, Alaska." In USGS Bulletin 849-E, 289–333. Washington, D.C.: GPO, 1933.

Rutledge, F. A. *Investigation of W. E. Dunkle Coal Mine, Chulitna District, Alaska.* U.S. Bureau of Mines Report of Investigations 4360. Washington, D.C.: GPO, 1948.

Sales, Reno H. "Ore Deposits of Butte, Montana." AIME *Transactions* 46 (1914): n.p.

Satterfield, Archie. *The Alaska Airlines Story.* Anchorage: Alaska Northwest Publishing Co., 1981.

Schrader, F. C. *A Reconnaissance of a Part of Prince William Sound and the Copper River District in 1898.* USGS annual report for the fiscal year ending 30 June 1899. U.S. Department of the Interior, 20th annual report, pt. 7. Washington, D.C.: GPO, 1900.

———. "The Contact Mining District, Nevada." In USGS Bulletin 847-A, 1–41. Washington, D.C.: GPO, 1935.

Schrader, F. C., and A. C. Spencer. *The Geology and Mineral Resources of a Portion of the Copper River District, Alaska.* USGS Special Report. Washington, D.C.: GPO, 1901.

Schwarz, Jordan A. *The Speculator: Bernard M. Baruch in Washington, 1917–1965.* Chapel Hill: University of North Carolina Press, 1981.

Shalkop, Robert L. *Eustace Ziegler: A Retrospective Exhibition, July 31–August 28, 1977.* Anchorage: Anchorage Historical and Fine Arts Museum, 1977.

Sheldon, Roberta. *The Heritage of Talkeetna.* Talkeetna, Alaska: Talkeetna Editions, 1995.

Sherwood, Morgan. "A North Pacific Bubble, 1902–1907." *Alaska History* 12, no. 1 (1997): 18–31.

Simpson, Ronald N. "The Bonanza-Motherlode Mine—A History." *Wrangell–St. Elias News* 5, no. 1 (1996): 7–8.

———. *Legacy of the Chief.* Anchorage: Publication Consultants, Inc., 2001.

Sinclair, Upton. *World's End.* New York: Viking Press, 1947.

Smith, P. S. "Mineral Industry of Alaska in 1930." In USGS Bulletin 836-A, 1–83. Washington, D.C.: GPO, 1933.

Smith, Sally Bedell. *In All His Glory: The Life of William S. Paley, the Legendary Tycoon, and His Brilliant Circle.* New York: Simon and Schuster, 1990.

Smith, V. Maurice. "Caught in the Riffles." *Jessen's Weekly,* 20 February 1958.

Spence, Clark C. *Mining Engineers and the American West: The Lace Boot Brigade, 1849–1933.* New Haven: Yale University Press, 1970; reprint, Moscow: University of Idaho Press, 1993.

———. *The Northern Gold Fleet: Twentieth-Century Gold Dredging in Alaska.* Urbana: University of Illinois Press, 1996.

Spencer, A. C. *The Juneau Gold Belt.* USGS Bulletin 287. Washington, D.C.: GPO, 1906.

Spude, R. L. S., and Sandra McDermott Faulkner, comps. *Cordova to Kennecott, Alaska.* Washington, D.C.: GPO, 1987; reprint, Cordova, Alaska: Cordova Historical Society, 1988.

Stegner, Wallace E. *Angle of Repose.* Garden City, N.Y.: Doubleday, 1971.

Stevens, Robert L. *Alaskan Aviation History.* Vol. 1, *1897–1928.* Des Moines, Wash.: Polynyas Press, 1990.

———. *Alaskan Aviation History.* Vol. 2, *1929–1930.* Des Moines, Wash.: Polynyas Press, 1990.

Stoll, W. M. *Hunting for Gold in Alaska's Talkeetna Mountains, 1897–1951.* Ligonier, Penn.: William L. Stoll, 1997.

Stone, David, and Brenda Stone. *Hard Rock Gold: The Story of the Great Mines That Were the Heartbeat of Juneau.* Juneau: Juneau Centennial Commission, 1980.

Thayer, Eliza T., and others. *Katherine Houk Talbott, 1864–1935.* Mount Vernon, N.Y.: Privately printed by Peter Beilenson, 1949.

Tower, Elizabeth A. *Icebound Empire: Industry and Politics on the Last Frontier 1898–1938.* Anchorage: Elizabeth A. Tower, 1996.

Tuck, Ralph. "The Moose Pass–Hope District, Kenai Peninsula, Alaska." In USGS Bulletin 849-I, 469–527. Washington, D.C.: GPO, 1933.

Tully, Jim. "The World and Mr. Nathan." *Esquire* 1 (January 1938): 42, 172–173.

Tweiten, Carl O. *Alaska, Big Delta and the Goodpaster Region.* Gig Harbor, Wash.: Carl O. Tweiten, 1990.

U.S. Department of the Interior. *Annual Report of the Alaska Road Commission.* Reports for fiscal years 1936, 1937, and 1940. Washington, D.C.: GPO, 1936, 1937, and 1940.

Van Hise, Charles R. "Some Principles Controlling the Deposition of Ores." AIME *Transactions* 30 (1901): 27–177.

Weed, W. H. "The Enrichment of Gold and Silver Veins." AIME *Transactions* 30 (1901): 424–448.

Who's Who in America. Various editions. Chicago: Marquis.

Who's Who in Engineering. Various editions. New York: John W. Leonard.

Who's Who in Engineering. Various editions. New York: Lewis Historical Publishing Co.

Who Was Who in America. Various editions. Chicago: Marquis.

Wickersham, James. Introduction to *Alaska: A History of Its Administration, Exploitation, and Industrial Development During Its First Half Century Under the Rule of the United States,* by Jeannette Paddock Nichols, 17–34. Cleveland: Arthur H. Clark Co., 1924.

Woodward, Kesler E. *Sydney Laurence, Painter of the North.* Seattle: University of Washington Press, 1990.

———. *Spirit of the North: The Art of Eustace Paul Ziegler.* Anchorage: Morris Communication, Anchorage Museum of History and Art, and Morris Museum of Art, 1998.

Zies, E. G., E. T. Allen, and H. E. Merwin. "Some Relations Involved in Secondary Copper Sulphide Enrichment." *Economic Geology* 11 (1916): 407–503.

Zimmerman, Henry A., ed. "Commencement Weekend Draws 500 Alumni." *Alumni News of Hobart and William Smith Colleges* 19, no. 4 (1955): 3–6.

Index

Page numbers in italics indicate illustrations.